Guide to Bees and Honey

Guide to Bees & Honey

Ted Hooper NDB

Blandford
Press

This book was designed and produced by
Alphabet and Image, Sherborne, Dorset
Line drawings by Elizabeth Winson

First published by Blandford Press Ltd 1976

Filmset and printed by BAS Printers Limited, Wallop, Hampshire
ISBN: 0 7137 0782 8

Contents

Acknowledgments

I would like to thank all those who have assisted in the production of this book, particularly Tony and Leslie Birks-Hay for their patience and great help in its editing and design, and those who have allowed me to use their photographs. The book could not have been produced without them. I would also like to acknowledge my indebtedness to David Rowse, on whose honey farm I worked for several years. The training in beekeeping I gained from him and the many discussions we had about the management of bees have formed the basis of much of my own attitude to the practical aspects of the subject.

Introduction

Many times in a year people of all kinds, and ages, come to me saying that they have always wanted to keep bees, and asking if it would be possible in their particular circumstances. The honeybee has an enormous fascination and many people get carried on a tide of interest into a relationship with this marvellous insect which is so unlike ourselves and yet lives in cities of 60,000 strong, without our need for committees and councils.

It is possible to keep bees anywhere. Some places will be more difficult than others and some places will be more rewarding than others in terms of the return in honey to be harvested. But the only real objection to keeping bees anywhere is that other humans may be inconvenienced by them, and in truth this is rarely the case—the concern people feel is usually not founded on fact but on ignorance. And if quiet strains of honeybee are kept and treated with reasonable care they are a problem to no one. Nor is it true that they will starve in the city or its environs. Often the amount of forage present in the form of trees in streets, parks, school and hospital grounds and the gardens of large houses provide a food density greater than that found around the country cottage in an arable cereal-growing area.

The important thing is that the prospective beekeeper should make some attempt to get to know what the handling of bees is all about before he purchases his first colonies. In this book I have tried to give the beginner the information he will require to start up and to work his colonies. The book does not set out to be an exhaustive work on honeybees or beekeeping; it merely tries to put over my own philosophy of beekeeping and the practical methods and approaches that I have found most productive in my own experience in a way which I have found to be helpful to my own students—if they will forgive me giving them this title—in Essex in south-east England.

The last few years have seen a considerable revival in interest in keeping bees, and it has also shown a trend towards keeping more colonies. The number of people with over forty colonies has risen and the number looking at part- or even full-time commercial beekeeping as a possible way of life has also increased. In this book I have to some extent, therefore, tried to point the way towards the running of a enterprise larger than just one or two hives. If what is written smooths the paths of some future beekeepers I shall be delighted.

The book is divided into four sections. The first deals with the honeybee and the honeybee colony. I have not given individual references to the origins of the statements made as I feel in a book of this type it is more valuable to tell a continuous story. The references, however, originate from the great specialist books listed on p. 257.

Section 2 goes on to deal with hives and equipment and the strategy

of beekeeping. This latter is most important: it is absolutely essential
for the beekeeper to have a feeling of the on-going process in a bee
colony and to see his work as part of this process rather than as a
number of individual manipulations to make the colony conform to his
ideas. It is possible to break the work done on colonies into units which
can be applied when the all-over strategy requires it. These units are
the tactics of beekeeping and are described in detail in Section 3. I
hope the beginner will be able to follow the instructions and get
himself, and his colonies, out of troubles in the early years. Section 4
deals with the flowers the honeybee works, and with the removal,
composition, handling and preparation for sale of honeybee products.

I have not dealt with pollination in this book for two reasons. The
subject is too large and too complex for a book of this sort to cover
adequately (there are two books in the bibliography which cover the
subject completely). To provide bees to do pollination is easy enough,
but to get bees to pollinate where and when you want is quite another
story and the beekeeper working in this field needs to do his homework
and study the crops as well as the bees. My second reason for not
dealing with pollination is that I am sure that honey production from
nectar must always be the prime motive of the larger beekeeping
enterprise—beekeeping as a pollination service is not viable economically as an enterprise on its own.

There is certainly room for more honey production. There are vast
areas of bee forage unexploited each year and the nectar from these
wild flowers or cultivated crops is lost with little advantage to man or
nature. Taking honey from an area removes very little other than
carbon dioxide and water and is of the sort of level of exploitation of the
environment that most will be prepared to forgive.

When first seen through the glass of an observation hive a colony of
honey bees is a mass of movement. Nearly everyone, young and old,
experiences a feeling of wonder that such confusion can be so fruitful.
Close observation, however, by many people, both scientists and
beekeepers, has brought some understanding of the apparent chaos. It
has often contradicted and destroyed the semi-mystical explanations
of the past, replacing them with more understanding of the individuals
and their complex society. We have learnt over the last few years many
of the ways in which the honeybee society relates to its environment
and some ways in which the behaviour of the individual adult is
controlled by feed-back systems so that the result is usually to the
benefit of the society as a whole. These concepts only partially clarify
the understanding of the colony. There is always the time when the
colony appears to make a decision to act in a certain way. For instance,
we know something of the feed-back system which allows the bees to
produce new young queens, but once they have these ready to emerge
from their cells, how does the colony decide what to do about them?

The structure of the colony is such that it normally has only one queen, so should the old queen, and later some of the young ones, go off with a swarm? Should just one queen be allowed to emerge and take over from the old queen, keeping the colony in one piece? Or should the new queen be killed so that the colony carries on as before? How this type of decision is made we do not know, though many ideas are put forward. Some scientists consider that bees make reasoned decisions, and that a type of deductive process can be demonstrated in the bee's behaviour when the physical picture of the approach to the hive is altered while the bee is away in the field. Others will want to demonstrate a more direct stimulus-response type of mechanism, with no 'thought' involved, and will point to various behaviour patterns occurring at the time as triggers for the next piece of behaviour, but there can be no certainty as to whether the pieces of behaviour are cause or effect. Either point of view may be the myth of our time which future research and observation will destroy or clarify.

The need to understand the honeybee colony, to have some working model of its organization and behaviour patterns, so that some forecasting of these may be possible, is not just an academic exercise. For the beekeeper it is an essential part of his equipment, to help him to plan his policy of husbandry and to make immediate decisions when he is manipulating his colony. I do not believe it is possible to work colonies successfully by a series of rule-of-thumb decisions if there is no understanding of the basic reasons for the rules.

In the first section of this book I therefore deal with the basic natural history of the honeybee, and particularly the parts which are of vital importance to the practical beekeeper. Those, and I hope they will be many, who become fascinated by the natural history and organization of the honeybee and its society will find a bibliography at the end of the book which will allow them to go further into the literature past and present of our subject.

Many readers will be more familiar with some aspects of beekeeping than with others, and I hope that they will be able, by looking at the Contents section on page 5 and by using the index, to go straight to the section which they need. For the beginner I feel I must cover the ground fully, but each chapter has been designed as a section in itself, and can be read as such.

There has been no attempt to metricate measurements in this book because at the time of writing the decisions have not been made about metric figures to be used in beekeeping. If I take it upon myself to do conversions I am bound to create inappropriate measurements. Beekeepers wishing to work in metric units must multiply inches by 2.5 for centimetres, pounds by 0.45 for kilograms and pints by 0.57 for litres.

1 Introducing the honeybee

The honeybee colony consists of a queen, who is mother to the rest, and worker honeybees numbering about 10,000 in the winter and rising to some 50,000 or more in summer. In the summer this will include some 200–1,000 drones, or males, which are killed off at the end of summer by the workers so that in the normal colony drones will be absent in winter. In addition to these adult bees the colony will contain a variable number of the immature stages of the honeybee. These consist of eggs, larvae—pearly white legless maggots—and pupae. The numbers of these young stages will vary with the time of year. All the immature bees are housed in the cells of the honeycomb, each individual in a separate cell, and are collectively spoken of as *brood*.

Packed into other cells of the honeycomb will be pollen and honey, the food of the bees, forming a store which can be drawn upon or added to as the circumstances allow.

This whole unit comprises a colony which is regarded as normal only when all the different stages are present. If any are missing the colony is at risk, even though this may be the normal condition for the time of year. The reason for this will become more obvious as we delve further into the life of the colony.

The honeycomb is made of beeswax. This is secreted by the worker bees from eight small wax glands on the underside of the abdomen (see page 18). When wax is required the workers fill themselves with honey, and probably some pollen, and then by hanging up in clusters retain the heat produced by the metabolism of the honey in their muscles. The increased temperature and the amount of honey in the bees cause the wax glands to secrete. The wax pours into eight pockets beneath the glands, and here a chemical change occurs which solidifies it. The result is eight tiny translucent white cakes of wax. These are then removed from the wax pockets by the last pairs of legs and passed to the mouth where each is worked and manipulated in order to form it into comb, or passed on to other bees for use elsewhere. The wax is

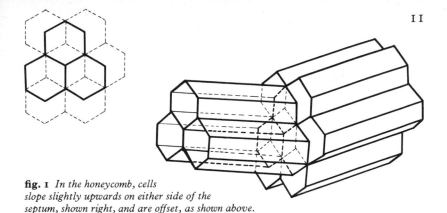

fig. 1 *In the honeycomb, cells slope slightly upwards on either side of the septum, shown right, and are offset, as shown above.*

moulded into position by the mandibles of the workers and the comb is quite swiftly built up to the size they require.

Honeycomb consists of hexagonal cells and is built up on both sides of a central vertical partition, the *septum*. The construction is shown in fig. 1. The base of a cell on one side of the septum makes up part of the bases of three cells on the other side. There are basically two sizes of hexagonal cell. Cells which are used to rear worker larvae measure about five to the linear inch and are called worker cells. Drone cells are larger, measuring approximately four to the inch. These are used, as their name indicates, to hold developing drone brood. Both kinds of cell are used for the storage of honey. The walls of the cells are extremely thin (about 0.006 inch) and are strengthened on the top by a coping, or thickening. When first fashioned the comb is opaque white with a rough, rather granular surface. It rapidly becomes creamy or yellow in colour as it is varnished and strengthened with *propolis*—the bee's glue obtained from plant buds—and brought to a high polish by the worker bees. When comb has contained brood, these areas become brown in colour due to the remains of cocoons and faeces left behind by passing generations. Comb gradually turns dark brown as time goes by, and old comb, though good, is almost black.

Honeycombs hang vertically and are arranged side by side. The number will vary in the wild colony, but in a normal hive there will be ten or eleven per horizontal compartment or box, spaced at $1\frac{3}{8}$ or $1\frac{1}{2}$ inch between septa. The space between the surfaces of the combs in the *brood area*—that occupied by eggs, larvae and pupae—is sufficient for two bees to work back to back. In the part of the comb where honey is stored the cells are extended so that the comb becomes thicker and the space is sufficient for only one layer of bees to work in it easily. The normal distribution of brood and honey in a comb is shown in the lower picture on page 52. Honey is always at the top of the comb and, if the brood area is small and honey plentiful, it may extend down the sides. The brood is below the honey, and pollen is usually stored in

worker cells in a band between the brood and the honey, but may also
be interspersed amongst the brood by some strains of bees.

Adult bees will cover the whole surface of the comb which is in use,
clustering densely in the brood area and more sparsely in the honey
store. These workers will be going about their various duties and will
at the same time be generating heat which will keep the temperature of
the colony up to the required level. This is about 17°C (62°F) when
there is no brood and about 34°C (93°F) when brood is present. This
heat is produced during the metabolism of honey to produce energy
for normal activities.

Having thus briefly described the honeybee colony we must look in
greater detail at the individuals. First of all I would like to look at the
adults, and the difference between the three types. Let us first examine
the worker honeybee, and then look at the way in which it differs from
the queen and the drone.

The body of the bee, like all insects, is divided into three main parts:
the head, the thorax and the abdomen, as shown on page 13. The head
carries a pair of feelers, or antennae, the mouthparts and the eyes. The
eyes are of two kinds: two large compound eyes which are the main
organs of vision and, on top of the head, three simple eyes, or *ocelli*,
which are probably monitors of light intensity. Inside the head is the
brain and several very important glands of which more will be said
later.

The thorax, or middle portion of the body, is divided into three
parts: the pro-, meso- and metathorax. Each of these segments carries
a pair of legs and the back two each have a pair of wings. The thorax
terminates in a segment called the propodeum, which is really the first
segment of the abdomen but which looks like an integral part of the
thorax. Internally the thorax contains the muscles of locomotion, the
largest of which are the huge muscles which power the wings and
which must be the main site of heat production both in flight and at
rest. These muscles are called indirect muscles because they are not
attached to the wings themselves but work by deforming the thorax,
the wings being worked with rather the same action as oars in a boat.
Small direct wing muscles deal with the feathering of the wing on each
stroke and control directional flight.

The abdomen is joined to the thorax by a narrow 'neck', the pedicle,
and is composed of six visible and 'telescopic' segments. Internally it
contains the alimentary canal, the wax glands, the heart, the sting and
its accessory glands in the worker and the queen, and the organs of
reproduction in both sexes.

The hard plates, and the soft membranous joints between them, on
the body of the bee are called collectively the exoskeleton. Unlike

*The drone in the centre of the picture, with big eyes, long wings
and stumpy abdomen, can clearly be distinguished from the
smaller worker bees on the comb.*

humans and other vertebrates, insects have their skeleton on the *outside* with the muscles internally attached. I often have the feeling that one or other of us must be constructed inside out. The exoskeleton is made up of two parts. The *epidermis* is a single layer of living cells which extends in a complete sheet over the whole of the body and lines the invaginations of the body such as the breathing tubes and the foregut and hindgut. Secondly, the non-living material secreted by the epidermis forms the hard, tough but flexible outer covering which we see as the outside of the insect, and which is called the *cuticle*. The cuticle is built of a structural substance called *chitin* (pronounced kitin), into which is injected a protein called *sclerotin*. This protein is tanned to form the hard plates but not in the flexible areas connecting the plates. The cuticle is not waterproof and the insect would quickly dry out and die if it were not for a very thin covering over the cuticle called the *epicuticle*. This is composed of several layers one of which contains waterproof wax protected from abrasion by a thin hard 'cement' layer.

The fact that the insect is covered by this 'dead' cuticle means that in order to grow it has to have a method of extending the size of its exoskeleton. The method which has evolved in insects is that periodically the entire cuticle is detached from the epidermis, which secretes a new cuticle inside the old one, the latter being mainly digested by enzymes which are secreted into the space between the new and old cuticle. Once these processes are completed the old skin splits and the insect wriggles out with its new, larger, very slack exoskeleton, which quickly hardens ready to start the next stage of growth. The whole process of getting rid of the old cuticle and growth of the new one is called *ecdysis* and the actual crawling out of the old skin is *moulting*.

Respiratory system

The breathing tubes mentioned above are called *trachea* and are the means whereby oxygen is conveyed directly to the places where it is required in the body of the insect. In all the 'higher' animals oxygen is carried to the tissues by the blood, but in insects the blood is not involved in the transport of oxygen through the body. The trachea are made of cuticle and are prevented from collapsing by a spiral thickening. The trachea start quite large but very rapidly divide many times, getting smaller all the while, until finally they end in single cells, or a loop. The trachea open to the air through holes in the cuticle called spiracles, and in many cases these are provided with a closing mechanism.

Air enters the tracheal system through the spiracles and fills the tubes. When the cells in which the trachea end are using up oxygen, this reduces the pressure of oxygen at that point and molecules of oxygen migrate in to make up the deficiency. It is thus by diffusion that oxygen makes its way via the trachea into the body of the bee. The oxygen is used to oxidize substances such as sugar in the cells to release energy for their use, producing the residue substances carbon dioxide and water. This is cellular respiration and is the reverse of the process *photosynthesis* whereby the plant manufactures sugar from carbon dioxide, water, and the energy of sunlight, allowing the plants eventually to secrete some of the sugar as nectar. In the honeybee, and many other flying insects, the main tracheal trunks become large sacs which are ventilated by the 'breathing' movements of the abdomen, whereby the abdomen is lengthened and shortened in a telescopic type of movement, and you can observe this movement in a bee at rest.

Circulatory system

As the blood is not involved in the carriage of oxygen it does not contain the red pigment haemoglobin and is a pale straw colour, or almost colourless. It contains many cells which are involved in such things as destroying bacteria, wound-healing, encapsulation of foreign bodies, and taking some toxic substances produced by metabolism out of circulation. The blood carries the substances resulting from the digestion of food around the body to the tissues and organs and also carries the waste products of metabolism back to the organs of excretion, the Malpighian tubules, for disposal. It also transports the hormones from the endocrine glands to the tissues which they affect.

The blood is not contained in tubes as in our own bodies but merely fills the entire space within the body, bathing all the organs. Circulation is accomplished by a 'heart' which is very unlike our own. It is found on the upper (dorsal) side of the abdomen in the bee, where it has five pairs of valves which allow the blood to enter when open, and extends through the thorax as a narrow tube with an open end behind the brain. A progressive wave of contraction runs along the heart, pushing the blood forward to be discharged in the head. This action causes a drop in blood pressure in the abdomen and increased pressure in the head thus causing the blood to flow backwards through the body cavity. This return flow is controlled by a number of membranes which ensure that the circulation reaches all parts of the body.

Alimentary system

Food is broken down by the process of digestion and these products are then circulated by the blood and used to provide energy, body-building substances, and the requirements for carrying out the chemical processes of life. The waste products of these processes have to be collected and eliminated from the insect's body. Digestion and excretion are the functions of the alimentary canal and its associated glands. These are shown in fig. 2. The mouth is between the base of

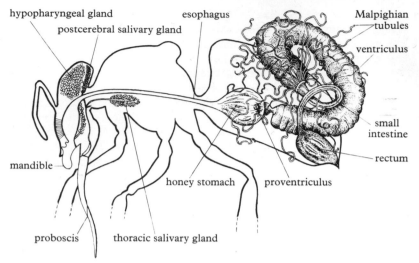

fig. 2 *The alimentary canal and associated glands of the worker.*

the mandibles below the labrum and above the labium. Immediately inside the mouth the canal expands into a cavity which has muscular attachments to the front of the head which can expand and contract it, thus providing some small amount of suction to help pass the food from the proboscis into the esophagus. Muscles in the esophagus provide waves of contractions which work the nectar back into the dilated crop or 'honey stomach', where it is stored for a while. At the end of the honey stomach is the proventriculus, a valve which prevents the nectar from going any further unless the bee requires some for its own use. If the bee is a forager it is in the honey stomach that it carries the nectar back to the hive, where it is regurgitated back into the mouth and fed to other bees. The proventriculus has four lips which are in continuous movement, sieving out solids from the nectar. The solids—pollen grains, spores, even bacteria—are removed from the nectar fairly quickly and passed back as a fairly dry lump, or bolus, into the ventriculus. When the bee needs to have sugar in its diet the whole proventriculus gapes open and an amount of nectar is allowed through into the ventriculus, where the food is subjected to the various enzymes which break it down into molecules small enough to be passed through the gut wall to the blood. The bee appears to digest only two main types of food, sugars and proteins. These are digested by enzymes produced in the walls of the ventriculus, assimilated and used to produce energy or to build up the bee's own proteins.

The residue is passed into the small intestine, and from there into the rectum where it is held, as faeces, until the bee is able to leave the hive and void the contents of the rectum in flight. During long spells of cold weather in the winter the rectum can extend almost the whole length of the abdomen before the bee is able to get out for a cleansing flight. At the end of the ventriculus are about a hundred small thin-walled tubes. These are the Malpighian tubules which have a similar function to our kidneys in that they remove nitrogenous waste (the results of the breakdown of proteins during metabolism) from the blood. The waste products, mainly in the form of uric acid, are passed into the gut to join the faeces in the rectum.

The alimentary canal of the larva is less complex than that of the adult. A very short foregut carries the food from the mouth to the midgut in which the food is digested. Up to the end of the larval period, that is until it has finished feeding, the midgut has no exit to the hindgut and the residue of food digested in the midgut remains there until the larva has finished feeding, thus preventing it fouling its food. When the larva is fully fed the hindgut breaks through into the midgut and the contents are evacuated into the cell. The four large Malpighian tubules, which had been removing waste from the body cavity of the larva and storing it, also break through and discharge their contents to mix with the faeces. The faeces are daubed around the cell walls and covered with the silken cocoon which is being spun by the larva at this time.

Glands of the head, thorax and abdomen

Just inside the mouth are the outlets of a pair of very large glands situated in the head and packed around the brain. These are the brood food, or *hypopharyngeal*, glands of the worker honeybee and these are of enormous importance in the life of the bee. The glands are composed of a large number of small spherical bodies clustering around a central canal. These bodies are made up of a number of secretory cells, and in the young bee they are plump and round. It is here that part of the brood food, a form of bee milk which is fed to the larvae, is produced. As the bee grows older and becomes a forager these round bodies of the gland become smaller and shrivelled: they are not producing brood food now but have changed to the production of the enzyme invertase, which inverts sugars. Should it be necessary for the survival of the colony the forager can, however, get this gland to produce brood food again and is thus able to feed larvae. The bee which has to survive the winter and who therefore must live longer than the summer bee has the gland in plump, brood-food-producing condition no matter what its age.

A preservative is added to the brood food, preventing its destruction by bacteria. This preservative is produced by a pair of glands which secrete their contents on to the inside of the mandibles to be mixed with the brood food as it is 'piped' out. (I use the word piped because the action always reminds me of a baker piping icing onto a cake.)

Other substances produced by the mandibular glands in the worker include heptanone which acts as an alarm scent to other bees. In the queen the glands are much bigger and produce fatty acids which we call 'queen substance', which is of great importance in the control of workers by the queen. Queen substance will be dealt with in more detail later.

Two salivary glands occur in the head and thorax, ending in common ducts one on each side of the tongue. Their watery secretion is used to dilute honey and to dissolve crystals of sugar, particularly at times when water is scarce.

As will be seen in fig. 3, four pairs of wax glands are situated upon the underside of the worker's abdomen on the anterior part of the last five segments, each gland being covered by the overlapping part of the segment ahead. Wax is secreted into these pockets as a fluid which rapidly solidifies to a small translucent white cake, probably by chemical action rather than by evaporation. A bee with wax plates in the wax pockets is shown below.

On the upper side of the abdomen, on the front of the last visible segment (segment 7) is a gland called the Nassenov gland. This gland produces a scent which, when the gland is exposed and air is fanned over it by the wings, spreads out from the bee as a rallying 'call' to other bees. It is used to help collect stragglers when there is a disturbance in the colony, and also at times to mark forage, mainly where a scent is

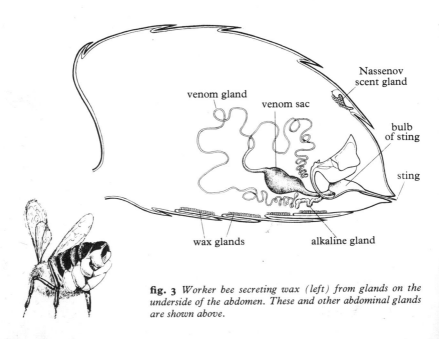

fig. 3 *Worker bee secreting wax (left) from glands on the underside of the abdomen. These and other abdominal glands are shown above.*

By turning down the last segment of the abdomen the worker exposes the Nassenov or scent-producing gland. The bee spreads the scent by fanning the air with its wings.

absent in the forage itself. The scent is not peculiar to the colony but is the same for all colonies as far as we know.

Finally there are the two glands associated with the sting. The long thin, bifurcated, acid or venom gland produces the venom which it empties into the venom sac where it is stored until required, and the short, stout alkaline gland is usually considered to produce a lubricant for the sting mechanism.

Nervous system

Every animal needs a mechanism which will allow it to test its environment and keep it from harm, or bring it to food and good conditions. In complex animals this job is done by the nervous system, and the actions of the animal are co-ordinated by the large collection of interconnected cells which we call the brain.

Insects have not only a brain in the head, but several smaller sub-brains or *ganglia* spread through the body. The larval honeybee shown in fig. 4 shows the brain and the string of ganglia running along the body on the lower, or ventral, side. Ganglia are more or less autonomous within their own segments but can be controlled and

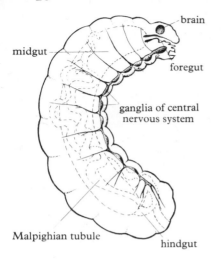

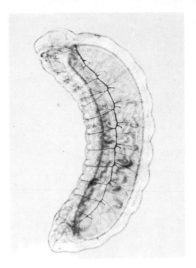

fig. 4 *(above) and the photograph (right) show the main organs of the larval honeybee. The additional dark branching line in the photograph is the tracheal system.*

overriden by messages from the brain. They also send messages back to the brain about the state of the environment in their area, thus providing the feed-back, and raw data, needed for the brain to function as co-ordinator.

We know little about the nervous functions and behaviour of the honeybee larva, mainly because it lives in a very stable and uneventful environment and needs to do little besides eat and grow. With the adult, we are dealing with one of the most advanced of insects, with an enormous repertoire of behaviour patterns and the need to check changes in its environment with considerable accuracy and blanket coverage. The brain of the bee is, in proportion to its size, very large. In the worker the brain consists mainly of the optic lobes, but the central portion contains the co-ordinating centres and this is larger, in proportion to the total size of the brain, than in most other insects. Two trunks pass from the brain around the esophagus to the ganglion below, from which another two trunks go back to connect with the first of two ganglia in the thorax, and then the five ganglia in the abdomen. Each ganglion has nerve fibres connecting it with the sensory endings on the outside of the insect, bringing data about the external environment, and others bringing information about the state of the internal organs of the body. Other fibres carry nervous impulses from the ganglia to the muscles and internal organs, regulating their action.

The sensory nerve endings, or receptors, are affected by changes in the physical and chemical environment and convert this information into electrical nerve impulses which can then be fed into the co-ordinating networks of the nervous system. The antennae are the main site for the senses, and other endings are found elsewhere over the bee's body.

The eyes of bees are totally different from our own. The main organs of vision are the two large compound eyes situated one on each side of the head, larger in the drone than the worker. Each eye is made up of thousands of tiny simple eyes, called ommatidia. Much argument has occurred over the years regarding what exactly an insect sees, and what sort of image is produced by each ommatidium and what is produced by the whole complex in the compound eye. It must remain a mystery in our present state of knowledge, but there are many things we do know about the vision of the honeybee. We know that it can recognize sights if suddenly taken out and released in country which it has already flown over. We can train it to come to various shapes to collect sugar and it can tell the difference between a square and a cross, though not between a square and a circle. The honeybee can see colour and differentiate between shades of at least some colours as well as we can,

The spoon-shaped mandibles, adapted for moulding wax, are agape as the worker sucks liquid through its proboscis. Just visible on top of the head is one of the three simple eyes.

The bee aims at the dark centre of the evening primrose (right). The dark marks are nectar guides which the bee can see because its eyes are sensitive to ultra-violet waves. Human eyes cannot see the nectar guides, and to us the flower appears as on the left.

although it sees different colours from those we see because its eyes are sensitive to a different part of the spectrum, being unable to detect red but detecting light in the ultra-violet region which is invisible to us. Finally, we know that its eyes are sensitive to the polarization of light, which we cannot see at all unless with the aid of certain crystals or polaroid plastic.

The honeybee is very well endowed with the senses with which it can monitor its environment and also with a large number of appropriate behaviour patterns which allow it to adapt to its environment over a very wide range of change. These abilities have allowed it to colonize the whole of the old world up to the arctic circle, and with the aid of man to extend its territory to cover the Americas and Australia.

Female reproductive system

Sexual reproduction helps to retain a large amount of variety within a species: children are never exactly like their parents, thus enabling the species to adapt to natural long-term changes in the environment. It does not, however, provide for the very rapid changes brought about by natural catastrophe or by the effect of man, and of his large all-pervading population, as is shown by the loss and diminution of many species of plant and animal in the last hundred years or so.

From the practical point of view the reproductive system of the queen bee should be well understood. It is illustrated in fig. 5. The abdomen of the queen is well filled with the two large ovaries, each of which is made up of over a hundred egg tubes or *ovarioles*. A single ovariole starts in the abdomen as an extremely fine tube which then widens, containing single large cells followed by a bunch of smaller ones. The large cells mature to become the eggs and the smaller cells, which provide the substances which build up the egg, wither away. The egg tubes on each side run into oviducts which then join to form the vagina, which opens above the sting. A large spherical sac called the *spermatheca* joins the vagina via a small tube. At the place of junction of this tube to the spermatheca it is also joined by the two tubes of the spermathecal gland. The sperms from the males with which the queen mates migrate into the spermatheca, probably chemically attracted, where they are stored during the whole life of the queen and fed by the secretion of the spermathecal gland. The workers have very small ovaries which, in the absence of a queen, can produce a few eggs. Workers which do this are known as *laying workers*, and the small egg-producing ovary from one of these is shown below.

fig. 5 *Much of the abdomen of the queen is taken up with the two large ovaries, as shown below.*

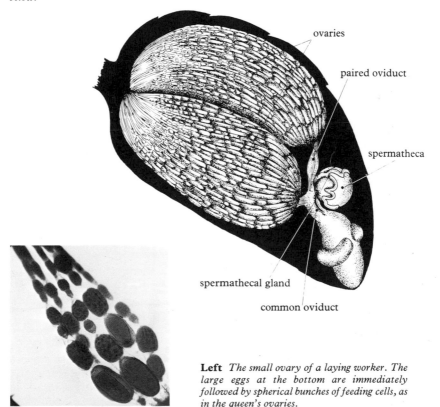

ovaries

paired oviduct

spermatheca

spermathecal gland

common oviduct

Left *The small ovary of a laying worker. The large eggs at the bottom are immediately followed by spherical bunches of feeding cells, as in the queen's ovaries.*

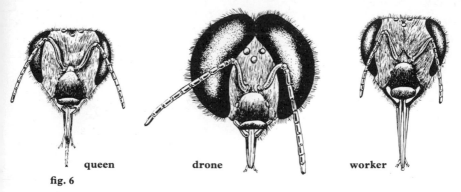

queen   drone   worker

fig. 6

Anatomical differences between the queen, worker and drone

The three different types of bee in the colony are called *castes*. The differences can easily be seen on pages 13 and 33 and in fig. 6. The queen is the longest of the three, her wings extending only about half way along her abdomen, which is pointed at the rear. For her size her head is proportionally smaller than the other two and she appears to be longer in the legs and more 'spidery'. The drone, which is about the same weight as the queen, is much more squarely built; his wings are very large and completely cover his abdomen, which is stumpy and almost square at the rear. His legs are long but his greater stoutness conceals this and he does not appear spidery. His head is large and almost spherical, being mainly composed of the two very large compound eyes which meet very broadly on the top of the head, reducing the 'face' to almost nothing. The worker is the smallest of the three, being about half their weight, and its wings do not quite cover the abdomen, which is pointed. Its head is proportionally quite large and triangular in shape and the legs fairly short. The worker is specially adapted to its work, and the biting mouth parts, or mandibles, are spoon shaped, without teeth, so that it can mould wax. Its third pair of legs are modified to carry pollen loads. The tongue is much longer than that of the other two castes, as only the worker forages amongst the flowers for nectar. The beekeeper will soon learn to recognize members of the three castes at a glance—a very necessary practical accomplishment.

Life cycle and metamorphosis

Having looked briefly at the anatomy and physiology of the honeybee we must now look at its development and at the origins of members of the three castes. The honeybee goes through four stages during its life cycle, these are the *egg*, the *larva*, the *pupa* and finally the *imago* or adult.

The eggs of the honeybee are *parthenogenetic*, that is they will develop whether they have been fertilized with a sperm from the male or not. All eggs, given the right physical environment, develop, and those which are unfertilized produce males, those which have been fertilized produce females. Drones, therefore, have no male parent and only one set of chromosomes, all of which come from their female parent. Females, on the other hand, are produced from fertilized eggs, having the usual double set of chromosomes; one set from each parent. As previously mentioned, the eggs are laid in three types of cell: drone cells (the large hexagonal cell), worker cells (the small hexagonal ones) and queen cells, which are much larger, thimble shaped, and hang down rather than lying horizontally. The queen lays unfertilized eggs in drone cells and fertilized ones in the other two: there is still argument as to exactly how she is able to do this but the best explanation from various pieces of research is that she measures the diameter of the cell with her front legs and if it is drone-cell size she lays an egg in it without letting any sperms escape from the spermatheca—hence the egg is unfertilized. A queen or worker cell will, however, cause her to allow sperms to escape into the vagina and the egg will be fertilized.

So much for sex determination. We are now left with the unusual problem of explaining the presence of two entirely different kinds of female in the colony. It can be shown quite easily that any fertilized egg in a normal colony will turn into either a worker or a queen, depending upon how it is housed and fed. The whole system of queen rearing is based upon this fact. Larvae taken from worker cells when they are very young and placed in cells hanging downwards are reared by the bees as queens. Therefore the genetic constitution of queen and workers is the same. There are no 'queen' eggs or 'worker' eggs; the difference is produced by a different method of feeding.

Honeybee eggs hatch in about three days whether fertilized or not. The tiny white legless larva is very soon surrounded with the white bee milk from the hypopharyngeal and mandibular glands of the nurse bees. If the larva is in a queen cell more and more of the white bee milk, called in this case royal jelly, is added until the larva is floating in a mass of food and eating to its fill all the time, up to and for a day after the cell is sealed over with its cap of wax. A worker larva is also fed a large quantity of bee milk, called in this case brood food, for the first three days, after which it is fed small quantities quite often. It is mass-provisioned for the first three days and then progressively fed up to the time the cell is sealed over on the eighth day.

Not only is there a definite difference in the quantity of food fed to the queen and worker larvae, the latter getting much less, there may also be a qualitative difference as well. No really consistent explanation of this qualitative difference has been demonstrated but it has been

After the egg (below left) hatches, the worker larva is surrounded with the brood food on which it feeds. During this period of growth and moulting it lies curled up at the base of the cell (below centre) until the cell is sealed. In order to take the picture of the larvae above the cell walls have been cut back, and it is possible to see the lack of bodily differentiation. When the cell is sealed eight days after the egg is laid the larvae turn sideways and become propupae inside cocoons, as shown below right. On the facing page can be seen three pupae of increasing age and, at the bottom, an imago ready to emerge.

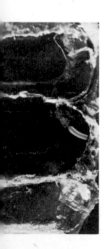  

shown that the rate of metabolism of the two types of larvae differs when they are only twelve hours old. This is long before there is a quantitative difference in the food. Also there are times when worker larvae are almost floated out of their cells on brood food but still develop into workers and not queens. So there may be a difference between brood food fed to the worker larvae and royal jelly fed to the queen. This difference may be due to different proportions of the output of the two glands involved in the production of bee milk, or perhaps to some additional 'hormonal' substance fed to the queen larva. It is certain, however, that the old idea that worker larvae are fed on pollen and honey only after the first three days is incorrect. Brood food is always the major part of their diet, although the amount of honey is increased after the third day and they may eat some pollen.

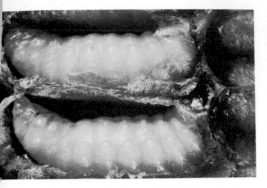

Domed cappings over drone propupae (left) can easily be seen in the lower part of the frame of brood (right). The empty drone cells around these are noticeably larger than worker cells.

During this period of feeding the queen larva has increased its weight by about 3,000 times and the worker larva by about 1,500 times. Extremely rapid growth on this very nutritious food can be made as there is very little need for digestion and very little undigestible residue. With this rate of growth four moults are necessary before the cell is sealed over and the fifth occurs after this has happened. It should be realized that in insects the larval stage is the growth stage; only at this time does the insect increase in size. In the case of the bee larva and most other insects growth is merely an increase in size, with very little difference between the anatomy of the large larva and the tiny one just hatched from the egg. The larvae are well packed with storage cells full of fat, proteins and carbohydrates so that when they are ready to pupate at the end of the feeding period they are at their greatest weight. After this their weight gradually reduces as some of the stored substances are used up to provide energy to build the adult body, which is quite a bit lighter than the fully-fed larva.

During the whole of the larval period the grub has remained curled up neatly in the bottom of the cell. When the cells are sealed over with wax the larva moves to lie down lengthwise in the cell. The bottom of the cell was carefully smoothed and polished before the egg was laid, but as the capping is put on from the outside, and the underside of it is quite rough, the larva responds to this surface by lying in the cell with its head outwards, against the rough surface. Before becoming quiescent, it defecates, spins its cocoon and then lies still, commencing the long change to adult. Approximate periods for the metamorphosis of the three castes are given opposite. There is little need to remember the whole of this unless you are entering for the beekeeping examinations, but the practical beekeeper must commit to memory the figures for the time taken from egg laying to hatching, from hatching to the time the cell is sealed, and the times of emergence of the adult insects. This is vital information on which a lot of practical work rests.

Little is known about the nutrition of the drone larvae, which are thought to be fed rather like the workers. They are produced in the larger cells and are sealed with a much higher domed capping than the workers, as shown opposite. The beekeeper should get to know the differences in capping as soon as possible because there are times when drones may be raised in worker cells, such as when there is a drone-laying queen or laying workers. The bees recognize the caste early and cap them with the high drone capping. Adult drones produced in worker cells will be much smaller than those produced in the normal drone cells, and these small drones, often called dwarf drones, spell disaster for the colony unless the problem is tackled and cured.

Days from Laying of Egg

	Worker	Queen	Drone
Hatching of egg	3	3	3
Larval first moult	$3\frac{1}{2}$	$3\frac{1}{2}$	4
Larval second moult	$4\frac{1}{2}$	$4\frac{1}{2}$	5
Larval third moult	$5\frac{1}{2}$	$5\frac{1}{2}$	6
Larval fourth moult	$6\frac{1}{2}$	$6\frac{1}{2}$	7
Cell sealed	8–9	8	10
Spinning cocoon	10	9	12
Propupa developing inside fifth larval skin			
Fifth moult to pupa	11	10	14
Pupa eyes pale red	14		
Pupa eyes red	15	12	
Pupa eyes purple, thorax yellow	17	13	
Pupa abdomen yellow	18	14	
Pupa antennae darken	19		
Moult of pupa to adult	20	15	$22\frac{1}{2}$
Emerges from cell	21	16	24
Mature and ready to mate	—	20	37

The above figures are averages and can be subject to variation owing to high temperatures.

Average length of life:

	Worker	Queen	Drone
Summer	38 days	3–4 years	22 days
Winter	6 months		

These are the mean lengths which are subject to wide variations both within and between the various races of bees in general use.

The emerging worker honeybee takes her first look at the world. The extreme hairiness of the face and eyes is obvious, and this hair will wear off rapidly within the first few days of 'rubbing shoulders' within the hive. The structure of the antennae is clearly visible: the long first joint, the 'scape', followed by the eleven joints of the 'flagellum'. The latter are covered by many sensory endings receiving and transmitting to the brain details of the environment, particularly scent, taste and touch.

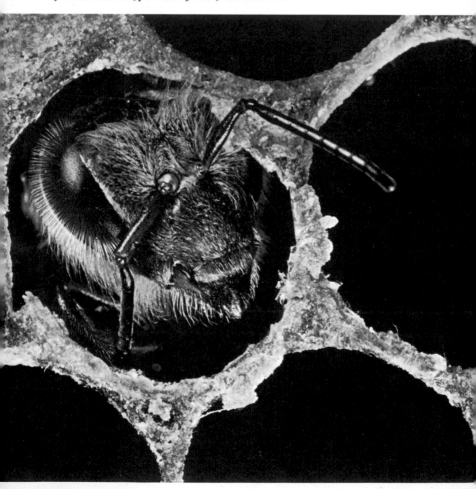

2 The bees behaviour

When the worker honeybee emerges from its cell, by biting around the capping and squeezing itself through the hole, it is slightly lighter in colour and more hairy than its older sisters. Young adults often take quite a while to get out of their cells because the other workers take little or no notice of them, and trample around on their heads quite happily. But eventually they draw themselves out of the cells rather like drawing a cork, and you might almost expect to hear a 'pop'. They are a bit staggery on their feet and of course quite useless to the colony; their glands are not working, they cannot sting, and the only work possible for them to do is cleaning up, which is just the job they do for the first three or four days. During this time they are being fed by the other bees and their glands become activated and ready for use so that by the end of the cleaning up periods their bodies are fully hardened and in full working order; indeed they will usually take a short, few minutes flight, not to forage for anything but just a trial flight around the hive: what the beekeepers call a 'play flight'.

Now the bee is ready to do any job required by the colony and, as observation has shown, it does a number of different jobs during the course of each day. In the normal colony there is a very loose progression through different jobs, according to age. This progression is cleaning, feeding larvae, manipulating wax, processing honey, and guard duty. However, any job can be done at any time and the needs of the colony are paramount. This indoor work continues for about the first twenty days of adult life, and during this time the beekeeper talks of the bee as a 'house bee' or a 'nurse bee'. After this it becomes a forager or 'flying bee' and its life style alters completely.

Behaviour of house bees

Many old grannies used to talk about the 'busy bee improving the shining hour'. I am afraid they were a long way out as far as I can see: the worker bee does not put in as many hours each day as most of us humans. The other thing is that it sadly lacks concentration and the ability to stick at a job for any length of time: observations have shown that it rarely stays working at a job for more than half an hour. Its day consists of resting, walking about and working in about equal proportions, but these are not done in large blocks of time but for a few minutes to half an hour at a time, and the three types of activity occur randomly. An observation hive gives one a chance to see all types of behaviour all the time. Resting bees often take up the most grotesque attitudes, with legs stuck out at queer angles and bodies jammed into corners, and one favourite place seems to be with their heads jammed between the top of the high domed drone cells and the glass. The bees which are on 'walk about' seem to do so with no particular aim but this activity includes several very important pieces of behaviour. Probably the most notable of these is the continual offering, and accepting, of food between individuals. This goes on to such an extent that every bee has very largely the same substances in its gut as every other bee. Not only is food passed around at this time but also chemicals known as 'pheromones', some of which are obtained by licking the queen and which control some of the behaviour patterns of the workers. As the bees walk around they also look into cells and come into contact with larvae that need feeding, ones which need to be sealed over, comb building, repairs, and all the work of the hive. It is possible that this continual recurring contact with the various needs of the colony will cause them to do the work that is required—a direct stimulus-response type of behavioural control. They also perform various dances which as yet mean little to us but must contain specific messages. Two of these dances are generally occurring at all times. One, where a worker does a sort of jerky, jitterbug dance on the spot, may be a request for grooming, as often another bee will dash over and start to nibble the dancer, particularly between the thorax and abdomen. Perhaps this is the bee equivalent of the spot between our shoulder blades: so difficult to scratch. In the second dance a bee rushes about all over the place, stopping occasionally to push its head against another bee, or to mount slightly on to its back, vigorously vibrating its body up and down all the while. It has been suggested that this has something to do with swarming, but I think it is too general in occurrence for this, and certainly does occur when there is no evidence of queen-cell production, either current or imminent. These 'walk about' periods of

Right *The queen surrounded by her retinue of workers, who are licking and grooming her. She is being fed by the central bee at the top of the picture while two other workers lean over to communicate with the queen, using their antennae. You can see the red tongue of the central bee on the right licking the abdomen of the queen and obtaining queen substance. The worker at top left shows the modified biting mouthparts, or mandibles, which are spoon-shaped in the worker so that they can manipulate wax. Eggs can be seen in some of the lower cells and three cells are sealed. The rather rough-edged cell between the two sealed cells at lower right is one from which a worker has recently emerged, and which the cleaning bees have not yet tidied up.*

the honeybee worker are probably very important in the general control and cohesion of the colony as a whole.

The work periods are short and interspersed with walking and resting, but with so many individuals work is progressing all the while, day and night. An army of cleaners is licking everything clean and polishing the inside and bases of the cells; damage to comb is being repaired and if necessary torn down and rebuilt; larvae have their food 'piped' in around them, and pollen is eaten so that the glands can manufacture bee milk. If I see larvae fully fed and ready for sealing in an observation hive during the day, by the next morning this sealing will have been done, and none will have been missed. The organization is extremely good but we have only hazy ideas about the mechanisms which trigger behaviour of this kind.

Every day there are batches of new adult workers emerging from their cells, perhaps 1,000–2,000 of them as the colony gets to its full size. These new workers find jobs in the centre of the brood nest and start their working sequence. The effect of this must be to push our original band of workers outside the brood area, where the work to be done is processing the nectar to honey and, for those who move into the area below the brood area, taking nectar from the foragers coming in from the field. The honey processors add the enzyme invertase to the nectar, and by manipulating it on the crook of the tongue (proboscis) so that it forms a drop, increase the area of nectar in contact with the air and thus the rate of evaporation. The drop is then sucked back into the honey stomach and a new drop regurgitated for evaporation. This evaporation is assisted by the hundreds of bees which are fanning the air with their wings, thus replacing damp air in the hive by a flow of dry air from outside. This is particularly noticeable in the evening, when a good nectar flow is in progress; hives will be roaring with the sound of fanning at dusk and the scent of the forage flowers will permeate the whole area. Much of these volatile scent substances must be lost from the nectar during processing to honey by the bee. The workers that receive from the foragers their loads of nectar also observe the dance the foragers do to communicate the position of the food they have been working, and at the end of their time as house bees these workers are directed to a particular flower and area of forage by this dance, which I describe in detail later.

From about the fourth or fifth day after emergence, adult worker bees go out on the play flights mentioned earlier. These tend to increase in frequency somewhat and time spent flying gets longer as

Left *A worker honeybee with a fair-sized load of willow pollen which she has collected, at least in part, from the big catkin on which she is standing. The pussy willow, or palm, is of great importance as a supplier of early pollen for brood rearing, and is well worth planting in or near an apiary.*

the bee gets older. Usually these flights occur in the afternoon, on a day with reasonably good weather: warm, dry, and not too windy. The young bees in an apiary often seem to go for play flights at roughly the same time: it is quite startling when you first see the apiary a-buzz with thousands of bees circling around. Many drones may be flying as well, their deeper buzz making an unusual note in the apiary. The beginner may feel a little intimidated at the number swirling around, but a few moments observation should reassure him, as the individual bees are taking no notice but fly round and round in circles of ever increasing radius. The foragers will of course be flying through this throng in their usual fairly straight lines, swinging down to the hive entrance and scuttling indoors as rapidly as possible. At the same time the outward-bound foragers are popping out of the entrance, often running up the hive front and away, pausing sometimes to wipe their eyes and antennae with the brushes on their front feet, and once airborne vanishing rapidly into the distance.

The play-flight youngsters, however, cluster in the entrance with quite a bit of grooming and mutual tapping of antennae. Once airborne they fly *backwards*, facing the hive but gradually circling away until finally they turn into their line of flight and circle around the hive, gradually spiralling outwards. During this performance they are learning to recognize the hive and the area in which it stands. It is as though they have an automatic cine camera inside them somewhere which photographs the hive, its surroundings and all the area they fly over. This photograph is remembered and the bee is able to use it to find its way back to the hive. It is quite interesting to test this out by removing a large stone or some object from in front of and near to the entrance to the hive. On arrival at the place where the picture has been altered many bees will remain fussing around this area for some while before going the last couple of feet to the hive. Some research workers feel that this period of delay and then the final flight forward demonstrates a basic deductive ability in the worker honeybee. Whatever the process, however, the bee certainly learns the look of the land for as far as it flies and hence the maxim, mentioned again later, that if bees are to be moved it must be under 3 feet or over 3 miles. Also, if for instance the grass is allowed to grow high in the apiary and is then cut down, chaos will reign for some hours or even days. The alteration of the picture over a large area will cause very considerable 'drifting', entry into the wrong hives, and the undoubted loss of some of the older foragers who cannot adjust to a completely changed visual environment.

Lines of workers fan their wings vigorously and draw air from the hive to keep the colony cool and evaporate water from nectar.

On guard, ready to challenge any intruder.

Guard bees

This mention of drifting leads us on to another facet of behaviour: guard duty and defence of the colony. In normal circumstances, when all is quiet and nothing troubling the colony, there will be no guards at the entrance. If, however, you tap on the hive a couple of times a bee will appear in the entrance; a few more taps will produce many bees at the entrance. Before long, if nothing can be seen of the cause of the disturbance, one or more will take to the air and have a look around to see what is going on. In other words, guards are only mounted when there appears to be a need, and this can arise from any new occurrence, like the tapping on the hive, by wasps or bees from other colonies trying to steal the honey from the hive, or animals rubbing against the hive, or even the vibration of a tractor or lawnmower nearby. I always remember an incident when a tractor driver left his tractor running opposite half a dozen colonies while he went to get something: we had to go and win it back from the bees for him because he did not fancy entering the milling crowd of bees that had surrounded it while he was gone.

Once guards are mounted they will run across and challenge bees entering the hive. What happens then depends upon the reaction of the other bee to the challenge. A forager belonging to the colony will completely ignore the guard and walk straight on into the hive, and the guard will recognize it as a colony member by its smell. This 'colony odour' is not an inherited scent (there is a slight genetic content) and this is obviously likely to be the case when it is realized that the workers of a colony are not all sisters but half-sisters in many cases, hence their genetic origins are not entirely the same. The colony odour is a product of the food eaten, and because of the very thorough food transfer all the bees in a colony have the same substances in the same proportion in their gut, and so their scent is the same. Different colonies will have different odours because they will have a different mixture of the various flowers of the district. Where colonies are on large areas of a single flower this differentiation by colony odour breaks down, much drifting occurs and often colonies become extremely irritable.

If a guard challenges another member of the colony, it recognizes its 'friendly' smell and does not press its challenge. However, a drifting bee entering the colony by mistake, perhaps because it has been blown down to the hive by a cross wind, or misled by a similarity of the approach picture, will be challenged. In this case the guard will press its challenge because the smell of this bee is not the right one. The drifter, because its instinct says it is in the right place, will not try to fight the guard but will submit. If the drifter is facing the guard it will offer food, which the guard will usually ignore. If the guard is attacking from the side, possibly hanging on to a wing or leg with its mandibles, the drifter will tuck its tail in and stand quiet, with its head tucked down, or it may rear on to its two back pairs of legs, extending its tongue and strop this with its front legs. These patterns of behaviour denote submission and the guard, although biting and pulling at its wings and legs, and climbing all over it, will do no real harm and certainly not attempt to sting. As with all bees, the guard's concentration period is short, and in a few seconds it gets tired of the whole affair and lets the drifter proceed, to be challenged several more times by other guards. Of course, after it has been roughed up in this

The visiting bee in the centre of the melée is stropping her tongue in submission.

way several times the smell of the guards will rub off onto it and it will become indistinguishable from others in the colony. This behaviour pattern is used by the practical beekeeper when he unites two colonies by the paper method (see page 163).

The position is quite different when the intruder is a robbing bee or a wasp. In these cases the intruder fights back when challenged or tears itself away and flies off. A fight may end with the death of one or both of the combatants. In these fights the sting is used and once a successful thrust has been made instant paralysis of the victim is certain. The guards are alerted to robbing bees and wasps by their flight, which is a characteristic zig-zag flight across the entrance, trying to find a way in without encountering the guards, readily observable by both the guards and the beekeeper. This flight pattern alerts the colony and prevents robbing from getting under way in many cases, and colonies which do not mount guards or react to robbers quickly are soon wiped out. Defence of the colony is necessary for its continued existence.

Foragers

We have now dealt with the life of the worker honeybee during the two to three week period when it is a house bee. Let us now look at the second half of its life. A bee's life in summer averages thirty to thirty-five days from the time it emerges from the cell to its death, usually in the field—it simply fails to make it home. Over-wintering bees, of course, have a much longer life span, but they are much less active. During the period when the bee is a forager it may be fetching into the hive one of four things which are collected by the bees. These are nectar, pollen, propolis and water. Nectar is sugar, water, and various other ingredients in very small quantities, collected from flowers and brought home in the honey stomach. Pollen, again collected from flowers, provides the protein, vitamins and trace elements for the bee diet. It is brought home as a load on the hind legs, as shown on page 34 and in the upper picture on page 52. Propolis is 'bee glue', used for glueing down anything loose in the hive, filling holes too small for the bees themselves to get through, and for varnishing and strengthening comb. It is brought home from flower buds on the hind legs like pollen but can be distinguished by its shiny appearance from the matt surface of the pollen load. Water is needed to dilute honey so that it can be used by the colony, and to cool the hive when temperatures are very high. These substances will be carried by most bees at some time during their lives and are usually carried one at a time, although some bees carry combined loads of nectar and pollen. There is a suggestion, however, that some bees are exclusively occupied in carrying propolis throughout their foraging lives.

All of these substances except nectar are collected to satisfy the colony's needs of the moment. Water is not stored at all, nor is there

Mouth to mouth food transfer. The worker on the left takes food from the one on the right.

any reserve of propolis. Pollen is stored, but not in vast quantities, and even where 'pollen clogged combs' are a worry to beekeepers much more pollen could be brought in and stored if the colony wished to do so. The collection of nectar is unlimited and colonies will go on collecting it as long as it is available and there is room to store it.

About 2 per cent of the bees reaching foraging age become scouts, which go out into the field and fly from flower to flower, of any species, and work them for nectar. They do several flights of this sort, finally selecting the flower that in their experience gives the highest sugar concentration and is sufficiently abundant to yield a full load easily and quickly. Once this flower has been chosen the scout bee becomes attached to that particular species for the rest of its life. It goes back to the colony and performs a dance indicating the position of the flowers it has been working, and gives out a sample of the nectar it has brought in. The remaining foragers will become attached to one or other of these dances and will go out to forage from the place indicated by the dance. If these young bees can find a load easily, they too will dance so that more bees will be attracted to forage a rewarding crop. Should there be sufficient bees on the flowers in a particular area to remove the nectar and thus make collecting a load difficult dancing ceases and so does recruitment to the crop.

Scouts and foragers coming back to the colony give their nectar to house bees, and this comprises a second selection process with regard to the value of the forage. If scouts or young foragers bring back, let us say, apple nectar with 25 per cent sugar whilst others are bringing in kale nectar at 40 per cent or dandelion at 50 per cent, it is obvious which bees will unload more easily. The house bees will accept the higher sugar concentrations more readily, and if there is enough of the higher nectar available then bees offering the lower value food may find it so difficult to get rid of their load that they will cease to collect it, and move to a more rewarding forage plant. This selection process means that the colony makes the best use of the forage within its district. This same type of selection is also known to work when a colony is fed sugar syrup, as the very high concentration of sugar diverts the foragers from collecting nectar and causes them to collect pollen instead. This behaviour has prompted the feeding of syrup in orchards in order to increase the efficiency of bees as pollinators. A remarkable change in this type of behaviour, for the good of the colony, has been seen in bees in the tropics. When water is needed to cool the hive by evaporation the bees accept the weakest solutions of sugar, or pure water, in preference to the highest as they would do in normal circumstances, and in some cases they have even been known to dance to indicate a water source.

The bee food-dances are well known and are illustrated below. When the forage to be indicated by the bee is up to 100 yards from the hive the bee dances as in fig. 7a. This dance says in effect, 'there is nectar or pollen [depending upon which the forager is collecting] close by: go and look for it.' It gives no indication of direction or any definite distance. The bee runs around on the vertical face of the comb, in the dark normally, following the path shown. Bees will be attracted to the dancer and will rush around after it trying to keep within antenna-touch. The whole dance is about $\frac{3}{4}$ inch across and will move its

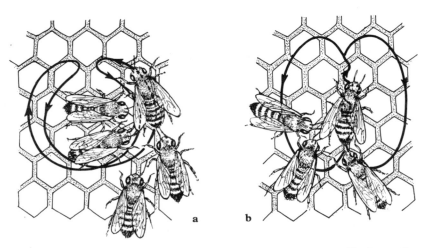

fig. 7a *A bee performs the round dance, indicating forage close by, followed by four workers who will later leave to search for the food.* **b** *The wagtail dance indicates both the direction and distance of the food.*

The worker in the centre is doing a vigorous wagtail dance. She has an unusually large audience following her movements and learning from them information on the food source. The more rewarding the source, the greater the vigour of the dance.

position as it is repeated. The bees which have followed the dance receive a sample of nectar, or will smell the pollen load on the dancing bee's legs and rush out to see if they can find the source. This is the dance which can initiate robbing in an apiary if honeycomb is left open to bees, or if honey is spilled on the ground. Once bees find honey (and in the latter part of the season it will only be a few moments before they will) they return to their colony with the spoils and dance, and within ten to fifteen minutes the whole area will be full of searching bees, trying to enter other colonies, and sheds and houses within a distance of 100 yards. The person who left the honey about will be far from popular!

When the distance to the food becomes greater the dance changes to the complete 'wagtail' dance at around 100 yards depending upon the race of bee involved. The wagtail dance is illustrated in fig. 7b. The dancing bee runs around the figure-of-eight and then across the centre in a straight line, wagging its abdomen vigorously from side to side. If the straight *wagtail* line points vertically up the comb it conveys the message, 'fly towards the sun'. If the bee runs down a vertical line the message is 'fly directly away from the sun'. By moving this straight

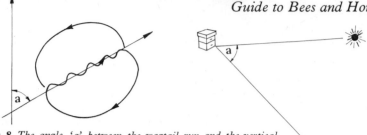

fig. 8 *The angle 'a' between the wagtail run and the vertical indicates the angle at the hive between the sun and the food source.*

run around at various angles (fig. 8) to the vertical the bee can indicate the same angles from the sun. On the vertical face of the comb in the hive it is using the force of gravity as its datum, and is indicating a direction in the field with the sun as its datum. The rate at which the bee dances, the number of complete figures-of-eight covered in a unit of time, the length of time spent on the straight wagtail run, and the duration and frequency of a buzz produced during the straight wagtail part of the run are all correlated to the distance to the food source being indicated. The longer the time spent on the wagtail run and the fewer the number of complete dances occurring in a unit of time, the further away is the source being indicated. The dancing bee stops every few seconds and hands out samples of nectar or allows followers to examine pollen loads on its legs. The followers are therefore provided with the direction to fly, the distance to the food, the type of food to seek, its scent and, in the case of nectar, its taste and sugar concentration. As far as we know, no information is given regarding the colour of the flower to be worked. The *modus operandi* of the followers is therefore assumed to be that they fly out in the direction indicated for the prescribed distance and then search coloured 'blobs' for the right smell, etc. Once they find the right flower they can then use its colour to reduce the time taken in searching for sufficient other flowers to obtain a load. There is evidence that the bee may continue, at least on one complete foraging trip, to work the colour of the first flowers it contacts. In some cases the bee can differentiate between shades which are quite difficult for us to separate and is seen to choose between two varieties of the same flower, working only one of them.

The dance described above is much more complex than this brief resumé indicates. The sun is moving all the while, but the bee makes allowance for this movement, adjusting its dance to the lapse of time between flying home and dancing. This shows it has a sense of the passage of time, a sense which has also been demonstrated by training bees to come to feeding stations at a particular time of day. The bee can also see polarized light from the sun, which we cannot. This means that the bee knows the position of the sun using the sky's pattern of polarization, which shifts relative to the position of the sun. This accounts for the bee's ability to obtain a directional angle for a sun almost at the zenith in the tropics. It also explains its ability to fly using sun compass in cloudy areas for only a little polarized light is enough to give it the information it requires.

Once the bee has become attached to a particular dance and subsequently to a particular patch of flowers of the same species, it tends to work these for the rest of its life. This is a short time only, about fifteen days, and so the flowers will often persist for this length of time. If, however, its particular species of flower comes to an end in the first few days of the bee's foraging, it will shift its allegiance to another plant, but should this happen towards the end of its life, it probably ceases to forage altogether.

In the field the bee works very economically, moving from one flower to another very close by. For instance, in orchards where the trees are planted in tight rows, with much bigger spaces between the rows than between the trees within the row, bees tend to work up and down the rows with very few crossing from one row to another. This is an important point to remember when pollination is required. In dense forage such as clover or crucifers grown for seed, or dense stands of heather, bees tend to walk rather than fly from one group of flowers to another. You may expect to see bees flying from one head of clover to the next, but if you look you will find them clambering about the maze of heads and leaves rather than in the air.

The queen and the drone

Although the queen will live for several years she lives a much simpler life than a worker. Her behaviour patterns are few and simple as far as we know, and she has not the wide abilities of the worker to deal with the general environment. The queen has evolved to the point where her only contact with the world outside the hive is during her mating flight and when she swarms, at which time she is well attended by workers. She has two main functions to perform in her life: mating and laying eggs.

The queen mates on the wing during the first ten to twenty days of her life. Once she has emerged from her queen cell she becomes mature within a couple of days, but by the time she is three weeks to a month old she is no longer capable of mating properly. During her mature period the worker bees become more and more aggressive towards her up to the time she mates. This behaviour has a possible value in driving the queen out for her mating flight before she is too old to accomplish it efficiently.

The drone's only function as far as we know is mating with the young queen, and as the drone dies when it mates any seen around have apparently served no purpose at all. However, I always feel they have much more use to the colony than we appreciate, since colonies

A drone being evicted by two workers.

which are denuded of drones never seem to handle normally, although it is hard to explain where the difference lies. Drones fly out of the hive when the weather is reasonably warm and fine, flying around at about thirty to ninety feet above the ground, depending upon the weather conditions. In some areas they form drone collections where they tend to congregate together in varying-sized centres of fairly high density. This seems to occur mainly in hilly and mountainous areas. In flatter areas with less erratic skylines they seem to spread out into a complete large low-density network over the whole area.

The queen on her mating flight flys up to the level of the drones for the day. Once she reaches the drone 'layer' the males, who have completely ignored her both within the hive and at other heights below the 'drone zone', are attracted to her by the scent of one of the substances (9 oxydecenoic acid) produced in her mandibular glands, and they form a 'comet tail' behind the queen and chase her by sight once they get within about three feet of her. The first drone to reach her is stimulated by another scent of unknown composition produced by the queen, which causes him to mate with her. At the time of mating the drone genitalia enters the queen and literally explodes, separating from the drone, which dies. The genitalia remain in the vaginal opening of the queen, with the torn end projecting from the tip of the queen's abdomen. This latter is what the beekeeper calls the 'mating sign' and is often seen on young queens who have just returned from a mating flight. The queen, however, does not mate with just one drone but average with about five to fifteen, on one, two or three separate flights. The 'mating sign' appears to be no hindrance to the queen's mating again almost immediately, and probably she can remove it with her legs. She appears to go on mating until her spermatheca is filled with sperms, as sperm counts are shown to be very similar in density despite variations in size of queen and spermatheca.

Mating having been accomplished, the queen starts egg-laying within a few days, and is from then on very carefully looked after by the worker bees. Up to the time of her mating they took little notice of her or were aggressive towards her, but now she produces a scent which causes them to turn and face her if she is close, thus forming the ring of workers usually found around the queen, and called her 'retinue' by the beekeeper (see page 33). These are not the same workers for very long, however, for as she moves around those she comes close to turn to face her but those at her rear are left behind and move away to continue with other jobs. While workers are in her retinue they lick her, clean her, and feed her, mainly on bee milk. The licking is very important because it is by doing this that they obtain various substances from the queen which control some of their behaviour. These substances, called pheromones, are described in more detail in the next chapter.

The queen lays her eggs in the bottom of the cells, a few a day in early spring but rising to a peak at the height of summer when she may lay 1,500 to over 3,000 in a day, depending upon the race and strain. This often amounts to far more than her own weight in eggs per day, and hence her need for large quantities of very easily-assimilated food. The need is supplied by bee milk from the bees of her ever-changing retinue, with some honey to provide the energy to keep her going.

Although the queen has this limited spectrum of behaviour—mating and laying eggs of the right kind in the correct cells—she is still by far the most important bee in the colony, both to the colony and to the beekeeper. She is mother of every bee in the hive. The whole inheritance of all members of her colony comes through her. This means that the working quality, the temper and the characteristics of the colony come from the queen. Change the queen and within a couple of months you have a completely new colony with, perhaps, quite a different temperament. Not only does the queen pass on her inherited characteristics to the colony, but also the number and the viability of her eggs will control the ultimate size to which her colony can grow. If she was poorly fed or came from a larva which started life as a worker and had been fed as a worker for a couple of days then she will never be able to produce a really good colony. Not only the inherited qualities of the queen but her own nurture and development will affect the future quality of the colony she will produce and head. The quality of the drones with which she mates will have considerable effect upon the inherited characteristics of her future colony, but the beekeeper finds this far more difficult to control. The control of mating is not yet a feasible technique for the vast majority of beekeepers; the practical difficulties are great and usually glossed over very rapidly by those advocating rigorous breeding. Breeding is a problem for specialists, who have not as yet made much advance in this area.

The importance of the queen to the colony and methods of obtaining good queens to replace her in due course should be something which exercises the minds of all beekeepers, particularly those who set out to obtain their living from honey production. This is dealt with in detail in Chapter 8.

3 The bee community

When dealing with a social insect it is necessary not only to look at the individual life and behaviour of members of the colony but also to look at the society as a whole and its behaviour as a unit. This is the way in which the beekeeper looks at his bees—in terms of colonies and colony behaviour rather than as collections of individuals and individual behaviour. The two do overlap and it is necessary to be aware of both.

In dealing briefly with development of individuals we have already dealt with several facets of colony behaviour, such as defence and foraging. Now I would like to expand what has been written in the two previous chapters in the light of colony organization.

The honeybee is thought to have originated in tropical areas and to have spread to other parts of the world by adaptation. The seasonal cycle varies in different parts of the world from the almost continuous round of flora in sub-tropical areas through two periods of fluctuation from dearth to plenty in the tropics to a definite annual peak and decline in the temperate zones.

In northern temperate lands there are little or no flowers producing forage for the honeybee from October to March, and it has to survive a six months dearth period using stores it has packed away in the previous period of plenty. Unfortunately the periods of plenty are often very short in countries like Britain and the whole crop is brought in by the bees in a short three-week period. The bee has adapted to this type of environment by a big cyclic variation in the size of population of the colony, synchronized with the availability of forage.

This annual cycle is illustrated by the annual population graph shown in fig. 9. This mean, or average, graph is much smoother than would be the case with an actual colony, which would show many short-term fluctuations in egg laying rate, especially in the early part of the season. I have shown the queen starting to lay in early January and from then gradually increasing her egg-laying rate. In my experience queens often start laying in December, have a short period of brood rearing and then shut down again. In the late springs of the early

1970's there was a tendency to delay the accelerating of the egg-laying rate until mid March, at which time the queens made extremely rapid broodnest expansions.

Keeping in mind such fluctuations, our graph gives us a good idea of the economy of a honeybee colony in the north temperate zone. In the early part of the season, through April to the beginning of May, the queen is increasing her egg-laying rate and the rate of increase is also accelerating, causing the very steep rise in the population of the brood. You will notice that at first this rises faster than the adult population so that ratio of adults to brood is approximately unity. This means that even in good weather the amount of forage which can be brought in by the adult population will be very largely used up in maintenance of the colony. This is partly because the proportion of the adult bees free from nursing duties and available for foraging will be quite small, and partly because forage in the early part of the season is of fairly poor quality, and the nectar low in sugar content.

As the generation of bees moves through its life cycle there is about a 50 per cent gain in numbers at the adult end because the worker bee is

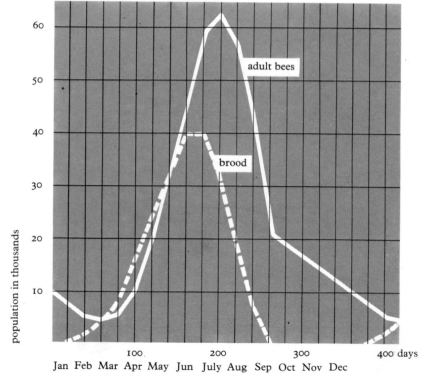

fig. 9 *The annual colony cycle shows three distinct periods in the ratio of brood to adults.*

twenty-one days in development and then lives for a further thirty to thirty-five days. By mid May the queen has completed her main increase in egg laying and the curve is now beginning to flatten out. This means that the ratio of brood to adult is nearer to $\frac{1}{2}$ than to unity and from this period on an increasing proportion of the adults will be foragers. This increasing foraging force will be servicing a broodnest which is ceasing to grow, and by the end of June is tending to decline in size, and hence the amount of food which will be required for colony maintenance will remain static, or fall whilst the amount coming in should be increasing. This is helped by the fact that the flora at this time of year is of much better quality, the clovers and crucifers having higher sugar concentration in their nectars than the spring flowers, and are on the whole more numerous over a given area.

The general tendency of the colony is therefore to build up its population using the output of the early flowers and then for this population to collect and lay in the large store of honey ready for the winter. By the end of July in many areas it is all over, and the brood population has been cut right back, which causes a rapid reduction in the adult population by mid August. This smaller population then lives on the stores through the winter, gradually diminishing until the following spring starts the increase in size once more. Without the help of a beekeeper the summer stores would have to be sufficient to last the colony through the winter and in many years a large number of colonies would starve. This is what did and still does happen where beekeepers—perhaps it would be more correct to call them bee owners—fail to look after their colonies adequately.

The annual cycle is the raw material that the beekeeper has to work on, assisting the rapid build-up of his colonies in the spring, holding them together during the period of peak brood-rearing when they may try to split up into swarms, keeping the brood-rearing going at the time when the queen is beginning to shut down if there is likely to be an August flow of nectar, and finally ensuring that they have sufficient stores to last them through the winter.

In other climatic zones the annual cycle is not as pronounced and the quiescent period does not last half the year as it does in temperate areas. A shorter quiescent period and a long foraging time gives heavier honey crops. In the tropics a double cycle may occur, with quiescent periods due to the rainy season at one end and to drought at the other. The basic beekeeping problems will be similar: the need to produce full-sized colonies, to prevent them breaking up into swarms

---

Right *A horizontal slice through a comb which contains eggs. The eggs are magnified approximately eighteen times, being about 1.8 mm long and 0.4 mm wide in actual size. It is most important for beekeepers to be able to see and recognize eggs in the cells, as much of the assessment of colonies rests upon the rate and number of eggs being produced.*

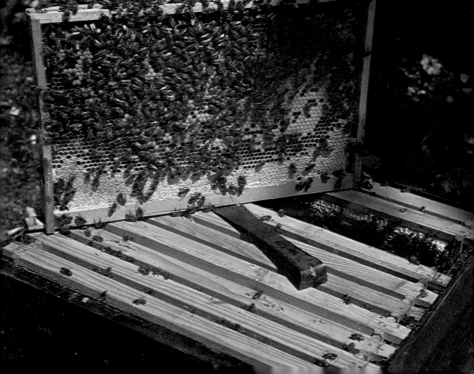

and to combat pests and disease. The sunnier areas have less trouble with the first two, but more with the last.

The phenomenon of swarming when the colony is at its peak population is well known, and I would like now to look at this, and its causes. The honeybee queen has evolved to a condition where she is capable only of laying eggs. She has lost entirely the ability to look after these eggs, to provide them with a home and defend them. All of these necessary jobs are vested in the workers. For the honeybee to reproduce its species it is therefore necessary to produce further queens who must be able to start a new colony somewhere else. The only way this can be done is for a queen to leave the hive with a band of workers to build and work for the new colony. In the wild condition this provides extra colonies so that those that are lost through accidents, adverse weather conditions, disease and predators may be replaced. For thousands of years swarming must have been the mechanism whereby the honeybee gradually spread out from the tropics and adapted itself to other regions.

A colony which changes its queen without swarming (known by beekeepers as 'supersedure') will be a new colony as soon as the workers of the old queen have died, and the whole population will then be the product of the new queen and will have different characteristics. This method fails, however, to increase the number of colonies and therefore does little to help the species to survive and nothing towards its spread.

Before either supersedure or swarming can take place one or more new queens have to be produced and got on to the wing. Queen cells are not present in the colony at all times however; they only appear when the time is ripe for supersedure or swarming to occur, or if the reigning queen is removed from the colony by the beekeeper. There must therefore be some trigger which initiates their production, and as a colony will usually show signs of the commencement of queen cells within twenty-four hours of the queen being removed, the trigger mechanism must react swiftly to her loss. The details of this

Left, above *A piece of comb showing stored pollen and nectar. There are at least two colours of pollen stored and two bees which have brought back loads from different sources are preparing to unload this into cells. The bee in the bottom left corner is taking food from a bee whose head is just in the picture. The straight lines running through some cells are the wires embedded in the wax foundation.*

Below *Hoffman-spaced frames in the brood chamber of a Langstroth hive. The wooden side bars keep the centres of the frames $1\frac{3}{8}$ inches apart. The comb which has been taken out, balanced on a well-designed hive tool, has been placed for stability upside down on the chamber. The white wrinkled cappings of sealed honey lie at the top of the frame in a gentle arch, and below it is the brown domed cappings containing worker pupae. The brood pattern is quite good, with some but not too many open cells mixed in with the sealed ones, and there is an average covering of worker bees.*

mechanism are as follows. We have already seen that the 'retinue' bees lick the queen and in doing so obtain from her body substances called pheromones. A pheromone, or ectohormone, is a substance produced by one individual which affects and alters the physiology, the behaviour, or both, of other individuals. The effect is obtained by very small quantities of the substance, which may be eaten or merely smelt by those it affects.

In this particular case we are dealing with a pheromone usually called 'queen substance' which is composed of at least two substances: 9 oxydecenoic acid and 9 hydroxydecenoic acid—the former being the same pheromone which acts as attractant to the drone when the unmated queen is on the wing. This substance is licked from the queen and passed around the colony by means of the normal food transfer mechanism. Workers who receive more than a very small threshold dose of queen substance in their food are inhibited from making queen cells. In the normal colony for most of the year this is the position. As the queen gets older her production of queen substance goes down, to about a quarter of her original production in her third year, but this reduction is in no way correlated with a reduction in egg laying. There will come a time, therefore, in some colonies when the queen is still laying a lot of eggs and building up a large force of worker bees but will not be producing sufficient queen substance to provide an adequate dose for all. The inhibition of some of the workers will thus cease and they will construct queen cells or allow existing incipient queen cups containing eggs to develop. The removal of inhibition is likely to be gradual, and possibly the production of incipient cups, the queen laying in them and some workers eating these eggs are all part of a gradual change away from inhibition.

A second way in which inhibition is thought to be removed is where the colony grows very rapidly, outgrowing its available room and becoming congested. In this case breakdown of the food transfer mechanism may allow some workers to become uninhibited and the result will be the same as above. The difference in this case is that it can happen to a colony with any age of queen. Congestion is one of the main causes of queen production and swarming.

Once the colony has started to produce queen cells it will then continue in one of three ways: it can swarm, supersede, or give the whole thing up, kill the contents of the queen cells, or young queens, and carry on as before. We do not know how the colony decides which path it will take; such knowledge could be of considerable importance in practical beekeeping if it were accompanied by easily recognizable behaviour patterns. Supersedure appears, from practical experience, to occur mainly in the autumn, during August. Often the new and old queens are found together, usually on the same comb. I would guess that some 5 per cent of colonies with 2 year old queens are in this state

A typical incipient queen cell cup built on comb overlapping the bottom bar of the frame.

each year, at least with the strains of bee that I have been concerned with. In Britain, swarming takes place mainly in May and June in the south, and up to three weeks later in the north. Some colonies which build up very rapidly in the spring, and colonies in areas where a very high density of early forage flowers occurs, may even swarm in April, and colonies slow to build up in some areas with no early forage may have their swarming period in July.

When the swarm leaves with the old queen only a portion of the colony goes with her and therefore, as she is producing the same amount of queen substance as before, the amount of pheromone available per bee will be greater and inhibition will return. The remains of the old colony and subsequent swarms will be headed by a new young queen who will be producing her maximum amount of queen substance and will easily keep the workers inhibited. With supersedure the colony will still be the same size as before but the new young queen will be producing considerably more pheromone, and normality will return to the colony.

The old queen which has gone with the first or prime swarm, which is the biggest in number, will build her new colony up as rapidly as possible, and it is possible that again she has insufficient queen substance to keep the rising numbers of workers inhibited. Thus, a certain proportion of such queens are superseded during the autumn of the same year.

The normal swarm or supersedure queen cells start as incipient cups which are laid in by the queen and then allowed to develop. They therefore start out right from their beginning as queen cells and are usually on the edge of or in holes in the brood combs, hanging downwards. When the beekeeper removes the queen from the colony,

or accidently kills her during a manipulation, queen substance ceases to enter the food transfer pool immediately and within a very short while the workers will start to make queen cells. As it is very unlikely there will already be queen cups with eggs in them, the bees make emergency-type queen cells. These are made by modifying ordinary worker cells containing worker larvae. The bees commence by adding royal jelly to the selected worker larvae until the larvae are floated up to the mouth of the cell. By this time the bees have modified the comb as illustrated above, shaping a queen cell from the worker cell of each of the selected larvae. The larvae are then floated into the normal position of a queen larva in the base of the queen cell. Providing the bees select a larva which is under thirty-six hours old the resulting queen may be quite acceptable. However, in their hurry they sometimes take older larvae, in which cases small queens, with fewer than the normal number of egg tubes, will result and these will be unsatisfactory as production queens to the beekeeper.

The queen substance pheromone has another effect upon the worker bee: it prevents the worker ovaries from developing and producing eggs. In the absence of a queen, and hence of queen substance, for some while the ovaries of workers do develop and produce eggs. These eggs are laid in the worker cells in a rather haphazard manner, the bees doing so being called 'laying workers' by the beekeeper. The worker honeybee is incapable of mating and therefore these eggs will be unfertilized but will develop and produce drones, dwarf in size because they have been produced in the smaller worker cells.

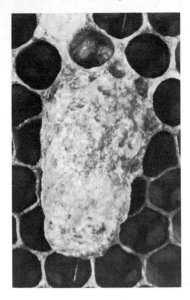

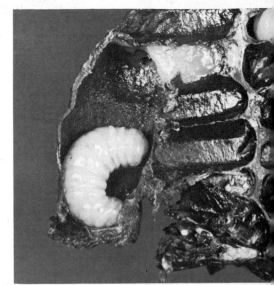

Queen cell cups are sometimes made in the centre of the comb, but the one shown left is an emergency cup made from a worker cell. A completed emergency cell is shown above sprouting from the comb, small in size and with a distorted cell next to it which may be an abortive attempt to produce another one. The picture above right shows the emergency cell in section, and its origin in one of the worker cells, which is still full of royal jelly.

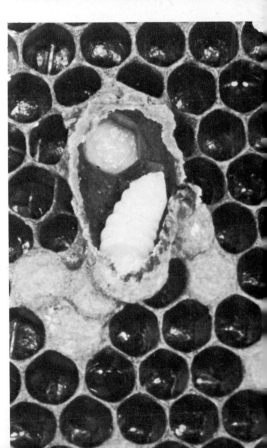

The picture on the right shows an emergency cell opened from the front. The worker larva was floated out of its normal position as the nurse bees added royal jelly until it reached the position of a queen larva. The queen larva goes on eating royal jelly for a day after the cell is sealed and here, as in the section shown above, the larva has eaten all the royal jelly in the base of the queen cell and some way into the worker cell.

True queen cells are at least half as large again as emergency cells. The section of a queen cell on the left shows a large queen pupa and, above her, a considerable residue of royal jelly. On the facing page four queen cells are shown in different stages. The top one has yet to be sealed. The second one has been cut open to show the pupa. The bottom one is intact, but between them is a cell from which the queen has gone, leaving the hinged cap attached.

One of the ingredients of queen substance, 9 hydroxydecenoic acid, is the pheromone which holds the swarm cluster together. The swarm comes out of the colony and usually hangs up fairly close by. If the queen is taken from it at this time the bees will return to their former colony. It has been shown that if the queen is taken away but the pheromone is placed in the cluster, on cotton wool, the bees do not break up and go home but are held together as though a queen were present.

We have already mentioned several other pheromones which help to control the behaviour of members of the colony for the benefit of all. Heptanone from the mandibular gland of the worker excites other workers' interest wherever it is deposited. Another, probably isoamylacetate, occurs in the venom or is produced at the time of stinging, and this calls other bees in to sting in the same place. The 'come and join us' scent from the Nassenov gland which calls in stragglers at times of upset and danger is a third, and is used at times to mark sources of food. There is a 'footprint' scent left by workers and the queen, marking trails over which they have walked and leading others to follow. There is also possibly one attached to drone cells, because it has been shown that the colony has an awareness of the amount of drone comb available to it at any time and uses this information to control the amount made anew.

There may be a pheromone which makes the presence of sealed queen cells known to the colony. I had an observation hive colony which had queen cells dotted about over an area 3 feet 6 inches across.

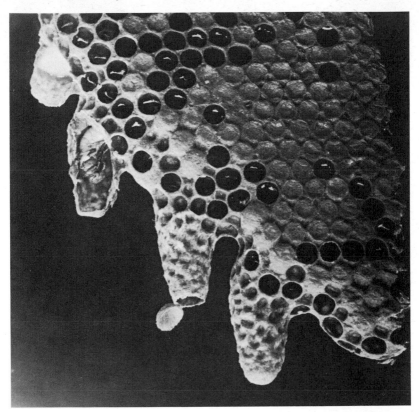

They swarmed several times, and as the population reduced the bees moved towards the entrance and abandoned a couple of queen cells which were on the extreme side, away from the main body. After a considerable bout of swarming only a handful of bees were left and these had no queen or queen cells available, the last having hatched and gone. This little cluster moved across the hive and sat on the previously abandoned and by now dead queen cells and stayed with them for about ten days, until I restablished the colony with a swarm. Something had called them over to the cells and in ten years of observing this hive I have never known the colony to leave the entrance except in this one instance.

Many other pheromones no doubt remain to be discovered in the world of the honeybee, as they are probably one of the main agencies of control in the insect colony.

We have looked at the way in which nectar and water are brought into the hive and passed around and stored. I would like to discuss this again in the context of the colony as a whole. I hope that fig. 10 will help. The large central block represents the nectar, or diluted honey, as carried in the honey stomachs of all the bees of the colony. Above this is the honey store to which honey will be added when in surplus or taken to replenish the honey stomachs of the central block. The contents of the honey stomachs may contain partly diluted honey and partly fresh nectar in proportions which vary with the nectar flow occurring at the time, and this will be used each day to feed the adults and the brood. The amount of their maintenance requirement will depend mainly upon the size of the broodnest and the number of feeding, or open, unsealed, larvae in it. The colony must have this maintenance ration each day to keep the brood alive. Coming into the central pool from below is the nectar being brought in by the foragers. The amount of this will vary with the acreage of forage plants yielding nectar within reach and the weather, which will affect the amount of flying the foragers are prepared to do as well as controlling the nectar secretion of the flowers. Finally the block on the left hand side represents the water-collecting bees, the number of which will vary with the colony's requirement for water to dilute stored honey.

fig. 10 *The honey–water–nectar complex.*

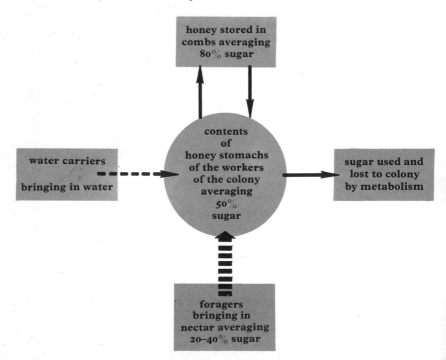

This is an ongoing process in every colony of honeybees all the time, day and night, summer and winter. Let us look at the summer first. In the early spring when no nectar is coming in, quite a large number of bees will be flying to the nearest water source and bringing it back to give to the bees in the broodnest. Some will be used by them to dilute stored honey, at the lower side of which there is always an uncapped band of honey, often partially diluted. The colony will be using up its maintenance ration each day and the size of the honey store will be diminishing. It will be used up from the lower side, and in a good colony the queen will be extending her laying into the cells as they are emptied. As the small spring nectar flows occur nectar will be brought in by the foragers and will reduce the rate at which the stored honey is being used up. Fresh nectar will, however, often stimulate the queen to accelerate her laying, and so the effect of small flows on the honey store may not be noticeable because of the increased maintenance requirement. Once a good nectar flow starts this will come in from the bottom and will make its way up through the bees to the honey store. Some honey will still be drawn from this store and water will be used to dilute it, but the number of water carriers will be greatly reduced. If more nectar is coming in than is required for the maintenance ration then the result will be a gain. It will be stored, and brought up to the correct specific gravity by fanning, to evaporate water. When the main flow starts, with a large population of foragers bringing in nectar as fast as they can collect it, the bees will no longer be able to hold the quantity in their honey stomachs and a considerable amount will be temporarily stored in empty cells in the broodnest, and sometimes even on top of eggs. Water carriers are now out of business as there is a surplus of their commodity in the hive. The maintenance ration is easily provided, and hundreds of bees will be processing the honey and passing it into the store where it will be sealed over as soon as the cells are full.

Bearing the above in mind, the beekeeper will realize that if the quantity of honey in the honey store plus the amount of nectar being brought in falls below the amount required for daily maintenance the colony dies, no matter what time of year this happens. He will also realize that, once he gets to know his colonies and where they obtain their water, the number of water carriers at work will give a very good idea of the state of the nectar flow, the number of water carriers being inversely proportional to the amount of nectar coming in.

The process goes on during the winter, but it is then interrelated with the process of temperature regulation much more than in the summer, and it is best to examine the colony from this angle.

The bee is a cold-blooded animal and tends to take up the temperature of its environment. Muscular action will raise the temperature of the muscles and the heat will spread through the body.

If the bee keeps flying, and thus keeps its body temperature up, it can fly around in temperatures below freezing. However, if it remains still and allows its temperature to fall to 8°C (46°F) then it will be immobilized for good. There is no temperature control mechanism in the individual bee's body but the honeybee colony, however, has such a mechanism and can control its internal temperature to within narrow limits over a very wide range of environmental temperatures.

As the environmental temperature falls below about 18°C (64°F) the bees begin to cluster together, forming a ball with the combs running through it. The top of the ball will be in contact with the store of honey and below this, where the combs are empty, the workers will creep into the cells, making the cluster almost solid. By the time the temperature falls to 13°C (55°F) the cluster is completely formed. The effect of the cluster is to reduce the heat lost from the bees. The bees in the centre eat honey and metabolize it by activity, thus producing heat, which can be lost by conduction, convection, and radiation. Losses by conduction will be insignificant, for both bees and wax are fairly poor conductors. Losses due to convection and radiation have been shown to be about equal and to be proportional to the surface area of the cluster. The loss of heat from the cluster can be controlled therefore by its expansion and contraction, and by coupling this with increased or decreased honey consumption the clustered colony has control of its temperature over a considerable range of ambient temperatures. In fact a cluster temperature of 31°C with an air temperature of −28°C has been recorded, a difference of 59°C (106°F). The temperature in the centre of the broodless cluster is kept at about 20°–30°C (68–86°F), which keeps the bees on the outside of the cluster on the bottom side, the coldest place, at about 9°C (48°F). Should the cluster cool so that these bees on the bottom fall to 8°C (46°F), they become immobilized, drop off the cluster and die.

Once brood rearing begins the brood area has to be maintained at temperatures of 32–36°C (90–97°F) or the larvae will die. The cluster is usually kept at about 34–35°C (93–95°F) when brood is present. However, larvae and pupae themselves produce a lot of heat whilst they are growing and undergoing metamorphosis, and will thus give some help to the adults.

During the winter period the honey being used will have to be diluted, and whenever possible bees will go out for water. They fly at quite low temperatures, load quickly and away. Often in the winter and early spring at about midday there will be no sign of flight and then suddenly twenty or thirty bees will return to a hive in a couple of minutes, then all will be quiet again. When the weather is too bad for even water carriers to fly the bees in the cluster dilute the honey with the output from the thoracic and postcerebral salivary glands. Water shortage is unlikely in the cluster as the metabolism of honey produces carbon dioxide and water as its main residues. The winter colony is helped considerably if the combs outside the cluster are full of honey because this acts as a heat reservoir and buffers rapid temperature change. Therefore the well-provided colony is doubly lucky: not only has it plenty of food within reach but is also helped in the control of temperature fluctuations.

Looking again at the graph of the population of a colony throughout the year (page 49), it is obvious that the bees which enter the winter are going to live considerably longer than the thirty-five days or so of their summer sisters. The winter bee is a rather different animal from the summer worker, the difference being brought about by feeding and by lack of work. In the late August and early September the workers feed very heavily upon pollen, and this brings their hypopharyngeal glands back into the plump form of the young nursing bee. At the same time a considerable amount of fat, protein and a storage carbohydrate called glycogen, or animal starch, is stored in the fat body. This fat body is an organ composed of a sheet of large storage cells spread along the inside of the dorsal part of the abdomen. It is present in all honeybees, but is considerably enlarged in the winter worker. It provides an internal store of food which is probably used to start brood rearing in the spring. These physical changes in the worker occur when it is not involved in rearing brood; in fact its lifespan appears to be inversely proportional to the amount of brood food produced and fed to larvae. In this way the lives of winter bees are extended so as to carry the colony through the winter, some of them living for as long as six months.

The same life-extending process comes into action when a colony becomes completely queenless. The last workers to emerge from the lost queen's eggs do not have to feed larvae, because there are none to feed, and they live for very considerable lengths of time. They go through the same anatomical changes as the winter bee: the fat body is enlarged, the hypopharyngeal glands return to nurse-bee condition and, in this case, because queen substance is missing, their ovaries also enlarge and produce eggs. By this means queenless colonies will go on living for the whole of a summer, but they rarely survive a winter.

We have now looked at the life cycle and behaviour of the individual, the annual cycle and the behaviour of the colony. This is the raw material which the beekeeper uses to decide upon what action to take in handling colonies in order to get them to produce a crop. It is absolutely essential to work *with* bees rather than to try and make them conform to your own ideas. You can *make* them do very little, and their objections are painful.

4 Getting started

This section is going to deal with practical beekeeping, and I shall begin with a subject which worries many beginners—bee stings. Everybody knows that the honeybee stings, but there are many old wives' tales about stings and how to treat them, and a short explanation might be helpful.

When the bee stings it injects a protein and various other chemical substances. There may be pain lasting about half a minute, but the main reaction occurs later, when the sting area swells as an allergic reaction to the foreign protein. The swelling may last, itching like a gnat bite, for a couple of days before disappearing. In my experience, this is the normal reaction of a good 90 per cent of people, and beekeepers gradually acquire a resistance to stings so that no swelling occurs after they have kept bees for a couple of seasons. The only substance which relieves stings is an antihistamine cream, but the majority of aspiring beekeepers do not need to use anything. Less than 10 per cent of people have a more serious reaction to stings, with the swelling increasing to alarming proportions or the development of urticaria, and in these cases medical advice should be sought. It is possible to have a course of treatment to desensitize oneself to stings. A tiny number of people suffer from hypersensitivity to the sting protein and become unconscious with ten minutes of being stung. They rapidly recover with treatment, however, but generally keep well away from bees thereafter.

When the bee stings it usually leaves its sting behind in the beekeeper, tearing off the end of its abdominal organs in the process and causing its own death within a couple of days. The sting will continue to pulsate, however, pumping venom from the venom sac into the wound. The quicker it is removed, therefore, the less venom will be injected. As the venom sac is attached to the part of the sting protruding from the wound, if you grasp it to pull it out you will

Suitably dressed beekeepers handling bees in a well-protected apiary.

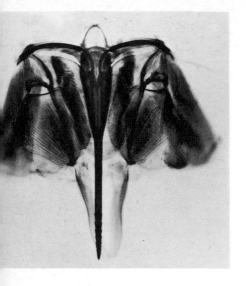

The·sting of the worker honeybee, dissected out and flattened. The sting is not a single needle but is composed of three parts. The barbs, which can be seen near the tip, are on two thin lancets which slide up and down on each side of the central body, actuated by the muscles and plates visible at the sides. Above the barbs the sting is thickened, and contains a pump which forces the venom through. The venom sac is not visible here, but can be seen on page 18.

squeeze all the venom into the wound. Instead, it is best to scrape the sting away with the edge of the hive tool or the fingernail, without compressing it. You are bound to get some stings, particularly when you first start, but with good-tempered bees and careful handling these should soon become very few and far between.

Personal equipment

The aspiring beekeeper should make sure that he has all the personal equipment he needs to handle his bees before they arrive. Essential equipment consists of veil, gloves, hive tool, smoker and overalls.

A veil is most important and should always be worn whenever one is handling bees. Why get stung on the face when it can easily be avoided? There are no prizes for getting stung, and it hurts. Neither does being stung make you a better beekeeper; so always wear a veil. There are many sorts of veil, both manufactured and home-made. An efficient veil should meet two criteria: it should be bee-proof—that is, the joint between the veil and the beekeeper should be—and the veiling should not blow against one's face in a wind—one's nose is in a very vulnerable position! I find it worthwhile to modify the bottom of my veils so that my neck is encircled by elastic in a hem in the veiling, and strings from the front are crossed and tied at the waist. To prevent being stung on the face in a wind the veiling can be held out on a hoop of wire or a wire box veil can be used.

I advise beginners to wear gloves because they will put a pair of gloved hands down to a colony of bees with much greater confidence than bare hands, and will keep them there more readily when bees land

fig. 11 *Separating frames with the hook of a hive tool.*

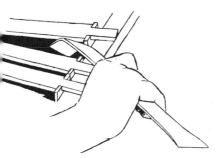

on them. This prevents a lot of stings in early days. But gloves are not just for beginners: bare hands soon get propolis on them and become sticky when one is handling bees, particularly in the warmer part of the season. It is then difficult to do delicate jobs such as handling queens. I prefer to wear gloves all the time so that when the need arises they can be removed to allow a pair of clean hands to do jobs like clipping the wings of the queen. I prefer the beekeeping gloves made of kid leather, with long gauntlets, but rubber and plastic gloves are satisfactorily used by many beekeepers, and cost far less.

A hive tool is necessary to lever the parts of the hive apart. Screwdrivers or old chisels should not be used as they will damage the hives, often leaving holes between the boxes where bees can get out, or wasps can get in. I prefer the flat broad-bladed type. The hook is used to prise the frames apart, as in fig. 11, and will do the job much more easily than the flat end. Considerably more leverage is available and frames are moved apart without sudden jarring, disturbing the bees.

The smoker is absolutely necessary and a good one should last a lifetime—particularly if made of copper. Two types of smoker are widely available: the straight-nose, or Bingham, smoker and the bent-nose smoker. The latter is the more efficient and will stay alight more readily than the former. A reasonably large smoker is a good investment. It is easier in use and does not need refuelling as often. Smokers placed on the ground often go out, and it is a good idea to screw a large hook, such as a coat hook, on the back of the bellows, so that the smoker can then be hooked on to the side of the hive where it is in easy reach as one is working.

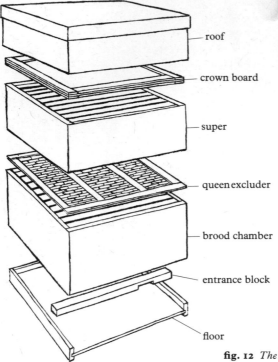

— roof

— crown board

— super

— queen excluder

— brood chamber

— entrance block

— floor

fig. 12 *The Modified Commercial hive.*

Overalls are not absolutely necessary, but bees get entangled in ordinary clothes like woolly sweaters, which never improves their tempers. White smooth-textured overalls are best. Blue cotton overalls should certainly not be used as they seem to excite the bees. This is probably the smell of dye or dressing used, as blue nylon does not have the same effect.

Hives

Bees have been kept for honey production quite successfully in earthenware pipes, straw skeps, wooden boxes and all types of hive. Given a cavity with a reasonable amount of room and protected from the main effects of inclement weather bees will manage, and will store honey if nectar-bearing plants are available. The different advantages of the various types of hive will be to the beekeeper, not the bee.

The modern beehive is made up of a series of square or oblong boxes without tops or bottoms set one above the other, with a simple floor at the bottom and a crown board at the top, and with a roof over all. Inside these boxes wooden frames are hung parallel to one another from ledges in the top of the sides. The bees are encouraged to make

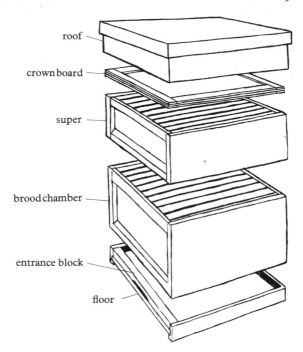

roof
crown board
super
brood chamber
entrance block
floor

fig. 13 *The Modified National hive.*

their comb within these frames, the beekeeper filling the frames with sheets of wax foundation on which the combs can be built by the bees (see page 74). An exploded diagram of the hive is shown in fig. 12, labelled with the names of the various parts. The only entrance to the hive is below the large bottom box, termed the 'brood chamber' because the queen is usually confined to this box, and hence it contains all the brood. The supers are used for the storage of honey, and the queen is prevented from going into them by the 'queen excluder', a grid of slotted zinc or wire with gaps large enough for the workers to move through, but too small for the queen.

There are four types of hive of this simple pattern in use. Arranged in ascending order of size these are the Smith, the Langstroth, the Modified Commercial and the Modified Dadant. To give you some idea of the difference in size the comb area available to the bees in each hive brood chamber is 2,186, 2,742, 3,020 and 3,805 square inches respectively. The British Modified National hive is slightly more complicated in construction, as shown in fig. 13. This is solely to accommodate the long lug of the British Standard frame. It contains the same area as the Smith British Standard frame using a short-lug.

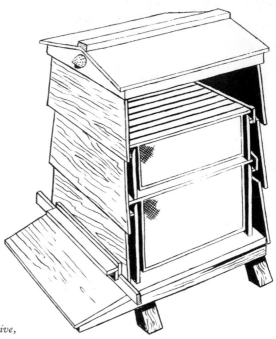

fig. 14 *The W.B.C. hive,*
lacking in efficiency.

The W.B.C. hive is even more complex, with an inner and outer series of boxes as shown in the diagram. It is a double-walled hive whereas all the others are single walled. It uses the British Standard long lug frame but holds one less than other hives so the actual comb area available to the bees is 1,988 square inches, which is in effect the smallest area in the bulkiest hive.

Hives	Number of frames in brood chamber	Capacity of brood chamber as a ratio; smallest hive = 1
W.B.C.	10	1
National	11	1.1
Smith	11	1.1
Langstroth	10	1.3
Modified Commercial	11	1.5
National and Super	22	1.7
Modified Dadant	11	1.8
Double Brood Chamber National	22	2.2

Frames	Length of top bar (in.)	Length of frame (in.)	Depth of frame (in.)	Effective comb area each side (sq. in.)
British Standard Brood	17	14	$8\frac{1}{2}$	93
British Standard Shallow	17	14	$5\frac{1}{2}$	55
British Standard short lug	$15\frac{1}{2}$	14	$8\frac{1}{2}$	93
Modified Commercial	$17\frac{1}{4}$	16	10	130
Langstroth Brood	$18\frac{3}{4}$	$17\frac{5}{8}$	$9\frac{1}{8}$	127
Langstroth Shallow	$18\frac{1}{4}$	$17\frac{5}{8}$	$5\frac{3}{8}$	66
Modified Dadant Brood	$18\frac{3}{4}$	$17\frac{5}{8}$	$11\frac{1}{4}$	159
Modified Dadant Shallow	$18\frac{3}{8}$	$17\frac{5}{8}$	$6\frac{1}{4}$	77

I make no apologies for dealing in this book only with the hives which I consider to be the most suitable in the light of my own experience. I have selected these hives because they are cheap, can be made easily by the do-it-yourself beekeeper, and are technically efficient. They are the very popular Modified National hive and the Modified Commercial hive. These two hives are to a large extent interchangeable. The exterior size varies by only $\frac{3}{16}$ inch and the main difference is in the depth of the brood chamber, as described above—2,186 square inches for the National and 3,020 square inches for the Commercial. This means that the beekeeper can meet all his requirements from one or a combination of these two hives. Beekeepers in countries other than Britain will have to substitute their own standard measurements for those which follow, as obviously one cannot deal with all forms in a single volume.

Whatever hives are used I would strongly advise that these are the 'top beeway' variety, most common in America, rather than 'bottom beeway' as used in most British hives. The difference is shown in the details in fig. 15. The quarter-inch space needed by the bee to move about between boxes is allowed at the *top* of boxes in the top beeway hive. In the other variety it is at the bottom of the box. Clearly a top beeway super must not be placed on top of a bottom beeway box, or

fig. 15 *Bottom beeway (left) and top beeway (right).*

there will be no space at all. Top beeway is much more efficient in use and less of a strain on the beekeeper as supers can be lifted back and placed 'cross cornered' on the hive and then slid around into place. With bottom beeway this cannot be done as the edge of the super box would run across level with the top of the frames and would decapitate any bee looking up between the frames and squash many of those walking about on top of the frames. The other advantage of top beeway is that crown boards and feeders do not need a beeway built on to their underside. Hives can be converted to top beeway by making the top rebate $\frac{15}{16}$ inch below the top instead of $\frac{11}{16}$ inch.

Frames

Each hive has its own frame, the National using the British Standard Brood frame in the brood chamber and the British Standard Shallow frame in the supers. The Modified Commercial uses the 16×10 inch brood frame and the 16×6 inch super frame. Of the frames on the market I would recommend the wedge-type top bar of $1\frac{1}{16}$ inch width, as shown in fig. 16, this width of bar reducing the amount of so-called 'brace' comb built by the bees in between the frames, to the inconvenience of the beekeeper. The frames illustrated are spaced by the extra thickness of wood at the top of the side bars. This is called 'Hoffman' spacing and is usually $1\frac{1}{2}$ or $1\frac{3}{8}$ inches centre to centre.

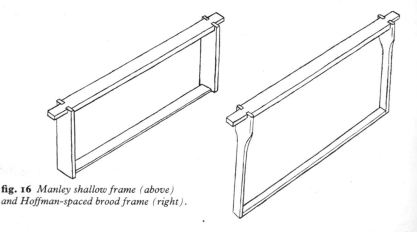

fig. 16 *Manley shallow frame (above)
and Hoffman-spaced brood frame (right).*

fig. 17 *Castellated runner in a super.*

Hoffman spacing is a great boon compared with other types of spacing and although it may cost you slightly more to buy your frames it is money well spent, and I would advise all beekeepers to use frames of the above type in the brood chamber.

Giving advice on which frames to use for supers is more difficult because the best frame to use depends upon the circumstances of the beekeeper, size of enterprise, etc. Hoffman *super* frames are a waste of money, and I find that the most efficient frame is the Manley spaced frame as illustrated in fig. 16. The frames are held quite rigidly and the spacing of $1\frac{3}{4}$ inch is designed so that the minimum number of frames is used consistent with a distance apart which is not too great when using foundation. Larger spacing than this will allow the bees to build their own comb in between the foundation, which they ignore. Another advantage of the Manley frame is that when uncapping the comb to extract the honey one rests the knife on the wood of the top and bottom bars, with a great saving of time. A problem with this frame, however, is that one needs to have a radial extractor for them (see Chapter 11), and these are costly. However, if the number of hives kept is likely to increase over ten or so, this type of extractor will help to speed up the harvesting anyway.

The beginner who feels he is likely to stay with less than four hives will probably buy, or hire, a small tangential extractor, and should therefore use the ordinary British Standard shallow frame and space them with castellated runners, as fig. 17, to provide nine frames to the super once the foundation has been drawn into comb.

Each of these frames should be provided with a full sheet of 'worker foundation'. This is a thin sheet of beeswax impressed with the hexagonal pattern of the honeycomb, and gives the bees encouragement to draw out the sides of the cells for brood or storage. Foundation can be bought in sheets and may be attached to the frame in two ways, either by using ready-wired foundation or by wiring the frame and then embedding the wire in the wax afterwards. Ready-wired foundation is the simplest and the process is illustrated in fig. 18. The frame is assembled, leaving out one half of the split bottom bar. The wedge is removed and the wired foundation slid down the grooves in the side bars until it fits tightly into the top bar. The wedge is jammed

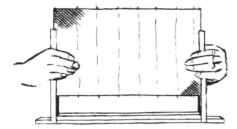

fig. 18 *The wired foundation is inserted in the sidebar grooves.*

The wedge is nailed in place.

in against the top of the foundation and nailed to the top bar. The best nails are called gimp pins and are available from the beekeeping suppliers. The second half of the bottom bar is put in and nailed to the *side* bars. On no account must the two halves of the bottom bar be nailed together. The sheet of foundation is fixed at the top by the pressure of the wedge and hangs suspended, the slots in the side bars holding it in place. It must be able to slide through the bottom bar when it stretches due to the heat and the weight of the bees working on it. The vertical wires prevent it stretching too much.

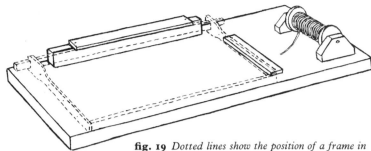

fig. 19 *Dotted lines show the position of a frame in place on the wiring board. A continuous wire is threaded in parallel lines and tensioned with the reel before tying off.*

I prefer to wire the frames because this means that the foundation and the comb, when it is drawn, is held centrally in the frame at several points. The method (below left) is to bore small holes in plain side bars, four in brood or two in super frames. These holes are protected by tiny brass eyelets to prevent the wire cutting into the wood. The wire is then passed through the holes and tightened to a good tension but not sufficient to bend the side bars. The wedge is removed, a plain sheet of wax foundation slightly smaller than the frame is placed behind the wires which are melted into the wax using about 9 volts to heat the

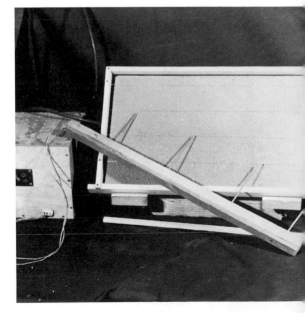

Left *A wooden jig for holding the frame sidebar when boring the holes. Eyelets placed on a sloping surface will all face downhill and can be picked up easily and thrust in the holes.*

Right *A sloping easel holds both frame and wax foundation in a convenient position for the electric 'prods' to be used to melt the wax on to the wire.*

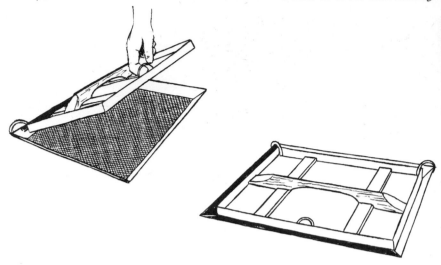

fig. 20 *A two-piece tray for making wax foundation.*

wire. Although this may sound quite a long job it is not once the necessary apparatus has been acquired. In any case I believe it is well worth the trouble as much more even, flat, comb results from this method than from using wired foundation. Beginners will probably buy ready-made foundation but after a good season should always have surplus wax sufficient for their own sheet requirements and for some small amount of increase.

To make foundation a piece of apparatus which will cast thin sheets of wax with the imprint of the worker cell on them is required. There are a number of these available both in Britain and Germany. They are easy to use and save money. The two plates should be lubricated with washing-up liquid so that each has a slight feeling of soapiness. Molten beeswax is poured on to the bottom plate and the top one lowered on to it. The surplus wax is poured off and the plates separated to reveal the sheets of foundation, which can be cut to the size required and the offcuts put back with the wax to be remelted.

I would advise the beginner to establish two colonies as soon as he can because often the problems of one colony can be sorted out if there is another one which can at times of need provide a frame of brood or even stores. Basic equipment is therefore two hives, each consisting of floor, entrance block, brood chamber, crown board and roof. A queen excluder will be necessary for each and I favour the short-slot zinc excluder which should be framed by nailing strips of wood $\frac{7}{8} \times \frac{5}{16}$ inch around the edges and two across the middle (see fig. 21). There is no

Zinc queen excluder with a reinforced wooden frame. **Fig. 21** *above shows the corners are bound with a thin strip of metal.*

need for joints, but the corners should be bound with metal strips. The beekeeper should aim to get an average of three supers per hive by his second season, so he will need six of these and their frames. One feeder, as shown on p. 130, per colony is ideal, as it is valuable to be able to feed all the colonies down at the same time in autumn—it is less likely to cause robbing than having some colonies feeding and some not. I recommend the Miller feeder. The final piece of equipment I would suggest is one or two 'nucleus boxes'. These can be made up by any amateur carpenter to hold about five combs each (see fig. 22).

Stocking the hive

Having looked at personal and apiary equipment, let us look at the bees to put in the hives. Bees vary a great deal. Some are good honey producers, some are poor; some are bad tempered, some are very placid. There are different races of bees. They may be rated as different varieties or different geographical sub-species. The ones we are likely to come into contact with are all one species, *Apis mellifera*, but will probably belong to or be hybrids between about four sub-species. *Apis mellifera mellifera* is the North European sub-species and *Apis mellifera ligustica* is the Italian sub-species. A mixture of these provides most of the 'blood' in the bees kept today. Italian bees have been imported into Britain in increasing numbers from about 1860. The North European bees are dark in colour whilst the Italian has a couple of yellow bands on the abdomen. Two other sub-species, *Apis mellifera carnica*, the

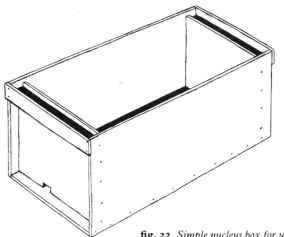

fig. 22 *Simple nucleus box for use in the summer.*

Carniolan race and *Apis mellifera caucasia*, the Caucasian race, have been bred into the beekeeping stock but in very much smaller proportions. In recent years queens which are of mixed race have been bred for honey production. These come mainly from the USA but one type, the Buckfast Abbey bee, is bred in Britain and multiplied for sale in the United States.

When beekeepers talk about the 'pure Italian', they mean a mated laying queen coming direct from Italy. The word 'pure' however leads to many misconceptions. The bee from Italy will be a member of the normal sub-species occurring in the Italian area, but it is also the result of many generations of selection by the professional queen breeders of Italy. As a result, if a number of queens is purchased from different sources they will show a great variation although individuals from the same source will be fairly constant. In other words there are many 'strains' of the Italian bee and none of them merits the title 'pure'. The same applies to any race or sub-species of honeybee; it will contain within it many strains.

In my experience there are some basic differences between the two main races. The Italian is good-tempered: the workers stay quiet and fairly still on the comb while being examined. They are good 'housekeepers', removing rubbish from the hive quickly and keeping everything clean, not tolerating intruders such as wax moth into the hives. They rarely kill their queens when these have been handled by the beekeeper for clipping or marking (see Chapter 7), and are more tolerant of examination during the period when they have an unmated queen in the hive, being less likely to kill her. The queens are large and yellow and can be found easily; they are prolific and tend to build up large colonies. Their main faults are that they are less hardy in cold winters than the darker races and during the brood-rearing season the queens tend to continue egg laying no matter how little food is available, and the colony can die of starvation.

The North European race is much more economical and tends to limit laying in times of shortage. The queen is rarely as prolific so that the colonies are smaller. The bees are more testy in temper and much more likely to 'ball' their queen (that is kill her by enclosing her in a small ball of living workers who hold her until she dies) if the colony is opened early in the year, or at times when the queen herself is handled. When the colony is opened and combs removed for examination the workers are given to rushing about making it difficult for the beekeeper to find the queen, who is already dark and therefore more difficult to distinguish from the workers.

We rarely, however, use 'pure' members of either of these races and it is with the hybrids that we have to deal. It is for this reason that the strain of bee is so very important. We do not want any of the bad characteristics but as many of the good as possible. Where winters are not too harsh and there is plenty of good summer bee forage, with the prospect of good weather at least in some years, then beekeepers favour the Italian or yellow strains. Conversely, in the areas where winters can be harsh and where even in the best of years nectar plants are not very plentiful, or if bee forage is plentiful the collection of nectar is likely to be heavily suppressed by bad weather, the dark North European race is more usually kept by the beekeepers.

My advice to the beginner is to get to know his local beekeepers, see what the majority of them do—particularly the successful ones who are respected by their fellows—and to try to obtain from them some local bees. This is particularly important where the dark bee is in general use, as sources of this type of bee are less well-organized than those selling the yellow strains, which are the major honey producers throughout the world.

Whatever honeybee you finally select it should have three characteristics: good temper, 'non-following' and stillness on the comb during manipulation. I would not tolerate the lack of these traits in any of my own bees. There is a saying in beekeeping that bad-tempered bees get more honey. This is not true. Good honey-producing strains can be quite calm and mild. 'Bad temper' will make your beekeeping less enjoyable and if you should become a full-time beekeeper then you will soon find that bad temper slows up the practice so much that bees of this sort are uneconomical. 'Following' is a separate characteristic from bad temper, and it is inherited separately. It consists of flying around the beekeeper after he has finished manipulating the colony and moved on. It is a trait which may make the beekeeper and his bees very unpopular with his neighbours.

The finger points to a queen cell, but this picture also shows bees running on the comb during handling, forming a cluster which will drop off the bottom bar.

'Running about' on the comb, often down the comb causing clusters of bees which then drop off the bottom bar, is another time-waster for the professional beekeeper and defeats the beginner in his efforts to find the queen.

The beekeeper should also strive to obtain in his bees 'good honey getting' and 'non swarming' characteristics, but these are more difficult to obtain and need not concern the beginner.

The beginner will usually obtain his bees in one of three ways. He may buy a full colony, a four or five frame nucleus, or get a swarm. A fourth method would be to buy package bees but this I do not recommend to the beginner as they require fairly careful treatment. Of the above sources I would recommend the purchase of a nucleus. Here you have a small complete colony consisting of just four or five frames of bees; not a very frightening sight even to the beginner. Bees, like other animals, are more likely to be difficult to handle as their number increases. The small colony will sit quietly and allow the beekeeper to deal with it easily. As the beekeeper builds the nucleus up into a full colony so will he build up his own experience and learn to handle them with confidence, so that by the time the colony reaches full size he is not worried by their large numbers. The new owner should, by buying from a reputable source, be assured of a docile strain of bee, and the nucleus should be a well-balanced colony containing a young laying queen, three or four frames of brood in all stages, a frame of honey and pollen, and sufficient worker bees to cover all this adequately even if the weather should turn cold. He should with luck be able to get a little honey—his first exciting crop—in the first year providing he receives the nucleus fairly early in the active season.

The beekeeper who starts with a full colony often does so because he wishes to have a full harvest as soon as possible. This is fine if he has a strong nerve, but in my experience the first time the beginner interviews a full colony on his own he finds it a very awesome sight, and this is not conducive to clear thought. Many mistakes are made because, with a full colony, he is often called upon to make major management decisions long before he feels at all at home with his bees. My advice is don't be hurried. Take your time; start with a nucleus. Get to know the 'feel' of the bees before you take on a full-sized stock. It is also much more costly to buy a complete colony.

The cheapest way of getting bees is to go and collect a stray swarm, either collecting it yourself or persuading a beekeeper to come with you and supervise your actions. This is fine, but you must be aware of the snags: firstly, you probably have no idea where they have come from and hence no idea of the type of bee and its handling behaviour; they can be quiet or very bad tempered. They are not usually bad tempered when you are taking the swarm but may become so when they are established. Secondly, they may be carrying disease, and if the brood disease American Foul Brood is confirmed, the stock and frames will have to be destroyed. Swarms are usually very free from disease but you should be aware that disappointment can result from this method of starting. If you are prepared to chance these two problems then go ahead—a bad-tempered colony can always be requeened from a good tempered strain (see Chapter 7).

Siting the apiary

Siting must be considered before the bees arrive because once they are in position and allowed to fly they can only be moved within the distances laid down by their behaviour. Bees always orientate back to the hive and if the hive is moved they will return to the place where they had learned it would be. This is so whether the hive is moved while they are out foraging or overnight while they are indoors. Obviously they do not orientate afresh each flight but rely on memory. Beekeepers have a rough practical rule which says bee colonies should be moved 'under 3 feet or over 3 miles'. With some strains of bee the foragers will come home to their old position and form a cluster and die there if their hive is much over 3 feet away. Three miles is twice the

Left *A well set up apiary in which five hives are accommodated in a small area of land in a garden. More room would be advantageous when examining the colonies.* **Right** *A Russian apiary in which the hives, on stands, are placed far apart, adding to the time taken over inspection.*

normal bee flight distance from a hive, so that from a new position more than 3 miles away they do not fly out and find their old flight lines and go down them to their old home. If they are shifted $2\frac{1}{2}$ miles then the new flight lines may overlap about $\frac{1}{2}$ mile and many bees will return to their last site.

There are two kinds of apiary, the 'home apiary' which is in the beekeeper's garden or, if he is a professional, at his main work base, and the 'out-apiary' set up away from the home. Out-apiaries are needed by the professional beekeeper and anyone with more than twenty or thirty colonies because the density of bee forage in most areas will not sustain such a large number of colonies in one place. Out-apiaries are also used by many beekeepers with small numbers of colonies because they prefer for many reasons to have the bees away from the home garden. This may be because someone in the family is allergic to them, either physically or psychologically, because they are hoping for better forage, or just because they want an excuse for a trip into the countryside.

At this point I will deal with siting and layout in general and details of the home apiary in particular and will leave the discussion of out-apiaries to later in the chapter. By 'siting' I mean where the apiary as a whole is to be set up. 'Layout' is the exact positioning of the hives in the apiary relative to one another and the topography of the ground.

Siting is governed by a number of general requirements which may be expressed as follows:

1 Easy access for the beekeeper.
2 Protective cover from the prevailing wind, or winds.
3 Good air drainage.

4 Away from heavy tree canopy.
5 Not overlooking or with main flight lines crossing public thoroughfares or footpaths.

Let us examine these rules one at a time with the usual house garden in mind. In this context easy access means with a hard path down to the apiary so it is possible to take equipment in and bring full supers out with the aid of a wheelbarrow or small truck. Try if possible to avoid having to climb over wire, down steps, or having to crawl under the low limbs of apple trees to get to the apiary.

Protective cover for the colonies is essential. Winter losses are usually higher in exposed sites than where good cover is present. Hedges are the best cover as the small amount of wind coming through a hedge prevents areas of turbulence which occur behind a wall. As winter cover is necessary a conifer hedge is better than one which casts its leaves. The site should be positioned so that the hives are protected from the main winds, but all-round cover is by far the best if it can be provided. For the site with no cover I would suggest the setting up of a temporary windbreak such as chestnut fencing while a hedge is planted. *Chamaecyparis leylandii* is a good fast-growing conifer for this purpose. This is the ideal, and if you fall far short in the possibilities you have available do not worry, as bees are kept in some very unlikely places with good results. Get as near as you can to the ideal but do not despair before you have tried a site out with a colony or two of bees.

Good air drainage is very valuable as it helps keep the site dry and allows cold air to flow clear of the colonies. If possible, keep apiaries away from the bottom of a dip in the land as it is likely to be a frost pocket and therefore will be a few degrees lower in temperature than

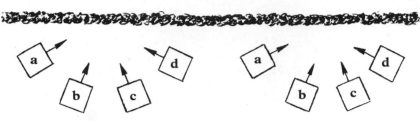

fig. 23 *A tidy layout, but one which will cause drifting.*

its surroundings. For the same reason keep away from walls halfway down a slope as the cold air will roll down and lie behind such places. Look around and try to find a place where cold air rolls on and away.

A heavy canopy of trees is rather like a frost pocket. It keeps the area underneath colder and wetter than a surrounding area which is clear of canopy. A really heavy canopy of trees such as in an orchard, small dense wood, or even a high very heavy hedge is bad. A site amongst trees which are well apart or in a clearing in woods, except where the trees are too high or too extensive, is often satisfactory. There was an idea at one time that wind vibration in the roots disturbed the colonies: this is a fallacy. The idea that one should not site beside an active railway line or an aerodrome is also nonsense. Bees get used to things very quickly. I have experienced apiaries in all these places and could never see any effect at all on the bees. Even when jets were taking off and passing over at about 400 feet the effect on the bees was nil, although on the beekeeper it was freezing.

Let us now turn to rule 5. Bees returning to hive appear to be flying blind or at least flying so fast that by the time they see an unexpected object, human or animal, in their path it is too late to avoid it. In most cases they bounce off and continue on their way but should they become entangled in hair or woolly clothes the stinging impulse is released and they will sting. They may hit, bounce off and then come back to have a look at what they hit more closely with quite peaceful intentions. However, the non-beekeeper does not know this, or will not believe it, and takes a bat at the bee with his hands. The result is often that an erstwhile peaceful bee comes back like a boomerang to do its worst. A similar thing can happen when the beehive overlooks an area where people are moving about. The guards at the entrance see the movement and may come out to investigate. Usually they are quite peaceful but can be very persistent, flying and hovering two or three feet or less from one's face on and off for quite a considerable time. This gets frustrating even for the beekeeper, and the non-beekeeper takes action very quickly, with the usual dire result. The moral is therefore not to site beehives where the bees can sit and look at people moving about, particularly non-beekeepers. This can of course be

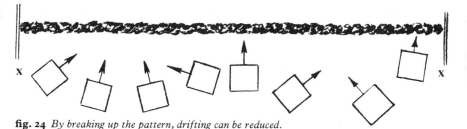

fig. 24 *By breaking up the pattern, drifting can be reduced.*

prevented by the hedge which is given to the hives for protection. Contact with flight lines is a bit more difficult and, if the public is inevitably in close proximity to the colonies, trouble can usually be avoided by using a high fence or hedge to push the bees quickly up into the air above head-level. Worker bees fly at about 15 feet on calm days and if pushed up quickly by an obstacle they are usually no worry. On windy days, and particularly with large apiaries, they may skirt a hedge and all go through a gap or gateway in very considerable numbers. This usually occurs in the country where people do not worry so much about bees, but in built-up areas it may be possible to create a more convenient gap for them if their normal route creates a nuisance.

Apiary layout should involve two main considerations: the prevention of drifting and the convenience of the beekeeper when working the colonies.

Drifting is always a problem. It is a considerable factor in the spread of disease, is conducive to robbing, and can cause the loss of queens when they are flying for mating. Considerable work has been done on drifting and it has been shown that it is at a minimum when colonies are arranged in a circle, with all hives facing in slightly different directions. It is unusual to be able to arrange colonies in a circle but straight lines should be avoided and each colony or each pair should face in a different direction. This is easy for a small number of colonies but often becomes more difficult in the large apiary, particularly where cover is only available on one or two sides. It is then impossible to get all colonies facing different ways but one should try not to repeat the all-over 'picture' for more than one colony. In fig. 23 the colonies repeat the picture and drifting will occur between **a**'s bees or **b**'s bees and so on. In fig. 24 the approach to each colony is different and drifting will be minimized.

I would always face the hives into a hedge and keep them between 4 feet and 6 feet away from the hedge. This prevents the bees overlooking the rest of the area and the cover provides protection and calm air in front of the hives. This is most important because as bees slow up to land quite light winds will blow them down. In positions where there is a long windbreak, as in figs 23 and 24, it is valuable to

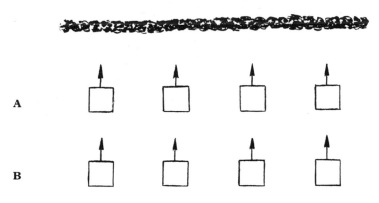

fig. 25

protect the ends of the apiary by short returns, as I have shown at X. These could be woven fencing, or if the apiary is permanent, a row of shrubs.

Another common layout consists of hives in more than one row, as shown in fig. 25, and is to be avoided on two counts: firstly, drifting will occur and secondly, as one works row A, row B will see the disturbance and will be alerted, making them more difficult to handle.

Many commercial beekeepers site their colonies in pairs, on stands. However, I would advise the beginner to site his hives singly, although he may still use double stands as he will find these useful at times.

I think stands are essential because if hives are low down beekeeping is a very backaching job, the beekeeper is not relaxed in stance and this makes him clumsy in his movements. I like the top of the brood chamber to be at the height of my closed hands when I am standing relaxed. This is the height of the normal dining room table: about 29 inches. Take about 11 inches away for the depth of the brood chamber and floor, and this puts the hive-stand top at 18 inches.

I use the normal commercial stands which are constructed as shown in fig. 26. Six pegs are driven into the ground for about 9 inches in normal soil, and the tops of the pegs are then levelled up by adjusting them accurately to the right height. The top rails are nailed to the pegs and the spacing pieces at the ends are nailed on to tie the whole together. The pegs and rails are made from 2 × 2 inch deal and the spacers 1½ × 1 inch. All the timber should be soaked in creosote for a while—the longer the better—and will then last a lifetime.

Distance between hives will vary with the type of apiary. The beginner can place his hives singly and have them about 6–8 feet apart. However, as the number of colonies in an apiary increases spacing will become more dense, unless there is unlimited room. Two factors

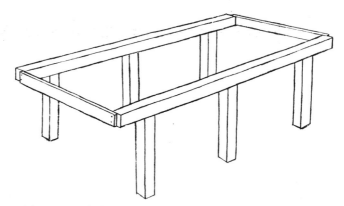

fig. 26 *A hive stand for accommodating two hives. The side rails are 4 feet 6 inches long.*

should be taken into account in every apiary. Firstly, ease of working around colonies and secondly, method of cutting grass. The same rules will apply whether we are discussing colonies arranged singly or in pairs on the stands described above, each pair being dealt with as a single unit. Enough space should be left between for the beekeeper to walk between them without difficulty while carrying a hive or supers. Particularly in the home apiary, which one tries to keep tidy, account should be taken of the type of grasscutter to be used and sufficient space given to allow cutting without a lot of shunting in awkward corners. Bees are never very tolerant of people cutting grass but if it is possible to keep moving onward all the time, they are less likely to be upset and to start following the mower.

Many beekeepers will want to expand and use out-apiaries. A permanent out-apiary will be sited and laid out in the way already described but whereas in the home garden the colonies have to be fitted willy nilly into that particular piece of land, when looking for an out-apiary the criteria of siting and layout can be used in the selection of the site.

In selecting a *district* in which to place an out-apiary try to assess the amount of bee forage plants available. Not only the main honey plants should be sought but also those which will provide both early and late supplies of pollen. These will help in providing good spring build-up and wintering (see Chapter 10). Take a look at the soils in the district and try to avoid the very light sand and gravel soils on which plants are likely to produce little nectar in drought years. Having decided that a particular district looks reasonable and is within the distance you are prepared to travel, start looking for sites and at the same time find out the position of any existing apiaries and keep at least three miles away from them. Consult beekeepers in the district to ensure you are not encroaching on their forage and to establish amicable relations for the future. When you have found a few possible sites, contact the owners of the ground for permission to put the bees there. In my experience it is best to go to a farmer with a definite request such as, 'May I put some bees in the copse over there?' rather than, 'May I put bees on your farm?' Farmers are usually quite amenable to having bees on their land but they are all busy people and the second question suggests that they will have to go on a tour of the farm with you to select a site, and it is often quicker to say 'No' if business presses. Traditionally the rent for an apiary is a pound pot of honey per hive per year, but the beekeeper must make his own arrangements.

When selecting the site it must be emphasized that easy access is very important. There must be a hard track right into the apiary, a track that will be passable to the beekeeper's vehicle even in the wettest year and which is not likely to be ploughed up or destroyed in the near future. Carrying beekeeping equipment into, and more particularly full supers out of, an apiary for any distance over rough ground is a very over-rated pastime and one of which I have had considerable experience. Out-apiaries should be concealed from the public as far as is possible. If they are easily seen then there is always the chance of vandalism. Little boys may go in and throw stones at the hives, older ones may turn them over or smash them up in other ways, and colonies may even be stolen. These problems are not very general, fortunately, but much trouble can be avoided by concealing the apiary or by putting it in a place that it is under the eyes of responsible people.

Damage to hives can of course be caused by animals. Horses, cattle and pigs will try to rub against them to relieve an itch and knock them over. All out-apiaries should therefore be fenced from animals. Three strands of barbed wire is usually sufficient, but ask the farmer's permission first, because some will not have barbed wire on the farm.

The only real assessment of out-apiaries is the amount of honey actually harvested from them. No one, by looking, can say what they will be worth. Even if an apiary is good this year it does not mean it will be so for ever: alteration in farming practice in the district can change it from good to bad in a single season. For this reason the beekeeper with a large number of colonies should always be trying out new sites and retaining the best. Extra sites not fully exploited provide for colonies from apiaries which must be abandoned quickly for some unexpected reason, and the more colonies managed, the greater the need for this sort of provision.

Temporary out-apiaries are used in migratory beekeeping and pollination work. Migratory beekeeping is where the beekeeper moves his colonies directly to fields of nectar-producing crops such as oil seed rape, mustard or clover, or to natural areas of dense forage such as heather or sea lavender. Site selection is not so stringent, and mobile

Hives placed on the ground are protected by the standing crop in this temporary out-apiary in Manitoba, but drifting would probably occur down wind.

stands are used rather than fixed stands. The most important requirement is cover. This can often be obtained from the crop itself if this is a high, dense one and colonies can be sited in the field. Where one is dealing with a crop plant which does not provide cover, such as clover or top fruit such as apples, a screen built around the hives helps

The nucleus has arrived in its travelling box with a deep screen on the top to prevent over-heating. It has been placed on its permanent site and opened up. The screen board should now be covered in case of rain.

both the bees and the rate of pollination. Straw bales, hessian or woven fencing can be erected as a temporary protection.

Stocking the apiary

The new beekeeper, having bought or made his hive, laid out his apiary and ordered his nucleus will eagerly await the great day when his own bees arrive. I hope by now he has also joined his local beekeeping association and has been along to some of their meetings and seen bees handled. In addition, if there is an Agricultural College near which offers courses in beekeeping, advantage can be taken of this to obtain a formal introduction to the subject. Every meeting one goes to increases knowledge and usually one gets some little practical tip on how to deal with the bees. Sometimes, let it be whispered, you see things which tell you how *not* to deal with the bees, and this can be equally valuable to the beginner.

The nucleus comes from the supplier in a well-ventilated box. It is well worthwhile, if it is coming by rail, to ask the station to let you know when it arrives so that you can fetch it quickly. The shorter the time bees are confined, the better. If the bees are buzzing loudly pour a cupful of water all over the top screen so that it will run inside. This will make them much more comfortable.

Having got the bees home take them down to the apiary, place the box on the stand they are to occupy and open the small entrance to allow them to fly. Remember this must be done only in the final position because all those who fly will orientate back to this position from now on. Obviously the beginner will put a veil on before he opens up the box to give himself confidence.

I would put a hive roof over the top of the box to reduce the light or rain going through the top screen. The little colony can now be left to settle down and explore its new environment until the evening. This allows them to get back to normal after their journey and makes them easier to handle.

In the evening, around 17.30 to 18.00 hours, the bees should be transferred to their hive. The beekeeper should first assemble what he will require near the nucleus: a floor, brood chamber, crown board, roof, two 'dummy boards' (solid pieces of wood the size of frames), three frames fitted with foundation, a supply of syrup and a feeder. The beekeeper will now don his protective clothing and get the smoker going (see page 119). A gentle puff of smoke drifting past the entrance of the nucleus box and in through the ventilation holes will tell the bees he is coming and they will follow their natural instinct and start to gorge themselves with honey. He should take his time and not hurry. He must give the bees a little while to get the message, then lift the nucleus box to one side of its present position. The hive should now be set up on the spot previously occupied by the nucleus. The floor is placed first with the entrance block in place, then the empty brood chamber in position on the floor, and the two dummy boards put into it, one against each outside wall. Another puff of smoke should now be drifted across the top screen of the nucleus box and the box opened by removing the cover or screen. Again a small amount of smoke drifted across the tops of the frames will send the bees down into the spaces between the comb. The frames containing the combs, with the bees on them, should now be lifted gently from the nucleus box and placed in the hive in *exactly the same order*, with the same faces of adjacent combs together. Care should be taken to see that the combs are not misplaced as they are transferred because comb faces are not flat, but tend to fit each other's hollows and bulges. If their position is transposed, bulges may meet bulges and bees may become trapped or even squashed between them. If one of these should be the queen the result will be disastrous for the development of the nucleus into a colony.

During the transfer of the combs, unless the weather is cold, the beekeeper may take a quick look at the brood stores and adult bees. Keep an eye open for the queen because it is very nice to know that she has been transferred safely, but do not set up a search for her because it is best to do the transfer without too much delay and with as little fuss as possible. In inclement weather do it as rapidly as possible consistent with gentle handling. Having transferred the frames with most of the bees still adhering to the comb, the few that are left in the box should

fig. 27 *To remove the rest of the bees from the box pick the box up in the right hand and bang it down on the left hand as shown.*

be knocked out over the top of the hive by picking up the box with one hand, inverting it over the hive and knocking it against the other hand (see above). The three frames of foundation should now be put in the hive. I would put all three frames of foundation on one side of the nucleus so that the arrangement in the hive would be dummy, frame of stores or smallest outside frame of brood, rest of nucleus, three frames of foundation, dummy. Many beekeepers like to put a frame of foundation on each side of the nucleus but I prefer to keep them together, which in turn keeps the clusters of comb-making bees together, on the basis that one big cluster is likely to be more efficient than two small ones.

For the beginner I feel there is considerable advantage in using a glass, or clear plastic, crown board to cover the bees. He will be eager to see what they are doing and will not be able to guess from the size and make up of the nucleus how fast they will pull out the foundation into comb and build up. This often leads to continual disturbances for the bees which will hold them back in their endeavours to grow to full size. A glass crown board allows the beekeeper to go to the hive, raise the roof and see quite a lot of what is happening without disturbing the bees at their work.

The feeder of syrup should be placed on top of the brood chamber to help the freshly-housed nucleus get on with the job. If a glass crown board is used and the beekeeper is to see into the brood chamber it will

The quart aluminium feeder; popular but inconveniently small for autumn feeding.

need to be a quart, round, feeder rather than the more practicable Miller feeder. Details on feeding are given on pages 130–33.

The roof should be put on and the little colony left to grow in size: it should not be opened up for at least a week. If a glass crown board is used the beekeeper need not open it again until he sees the bees beginning to work on the outside frame of foundation. Feeding should be kept up all the while. Usually, however, the new beekeeper is eager to see what is happening in the colony and so, glass crown board or not, he will want to look at the end of a week.

Opening up should be carried out as detailed on pages 119–22. Temperature requirements should be considered and the amount of time spent on manipulation adjusted to suit conditions at the time, remembering that this is even more important when dealing with a nucleus than a large mature colony.

The beginner must try to educate himself in the things he will find in a colony. He should look for eggs, larvae, sealed brood, honey both open and sealed, and pollen. He should note the position of these in the individual cells and their distribution in the colony as a whole.

Colour is another important factor to get to know; the colour of new comb as compared with old, the colour of cappings on these different-aged combs, and the difference in colour between brood and honey cappings (see the lower picture on page 52), the pearly white colour of larvae and the many colours of pollen. It is interesting to associate these later with the bee forage plants from which they are obtained—see Chapter 10. Dorothy Hodges' book *Pollen Loads of the Honeybee* is an invaluable guide to pollen colours.

You will not be able to remember everything at once but start right from the first examination with these things in mind because this information is all of importance in the management of your colonies.

The new beekeeper will be able to do little for the nucleus other than to give it more foundation as required and to keep it fed so that it always proceeds at the fastest rate possible. The more experienced beekeeper will be able to increase the colony's rate of growth by judiciously 'spreading the brood' as described later, but the beginner will possibly do more harm than good by interfering, especially when adverse beekeeping conditions are prevailing. The number of new frames required will depend upon the work the bees have already done on the first three frames of foundation. If they have 'pulled' or drawn out the comb on one frame of foundation and just started on the second, give them one more frame so that they have two in hand to work on. If they have pulled one and are well on with the second I would give them two more. If they have pulled two and started on the third then give them three more. Three will make up their full complement of frames and the dummies will have had to be removed. Further extension must now be accomplished by adding further boxes of foundation, usually in supers, rather than a few frames at a time. Rules for and methods of supering are given on page 128 and will apply because the nucleus has become a full colony, although still small in adult population. However, bees are often quite reluctant to move into a box of foundation through a queen excluder, particularly when the colony is still building up. To get over this it is a good plan to put the box of foundation on *without* an excluder. The bees will move up rapidly to pull comb and become established, whereupon the queen should be found and put into the brood chamber, if she is not there already, and the excluder placed in position between the brood chamber and super.

I am assuming that the beginner is starting with and intending to keep his bees in one brood chamber per hive, and that if he is in a warm area with good forage it will be a large brood chamber. In southern Britain I recommend the Modified Commercial and in a colder, more austere area the National. He may find difficulty in obtaining a Modified Commercial nucleus—if he does and finds he has to buy his bees on British Standard frames, he can put these in his Modified Commercial hive by use of an adaptor, as shown in fig. 28, to prevent the bees making 'wild' comb in the spaces. The frame in its adaptor can be handled in the normal way and removed from the hive when convenient—usually when there is no brood on it—and replaced with a sheet of foundation in a frame of the correct size, providing it is a time

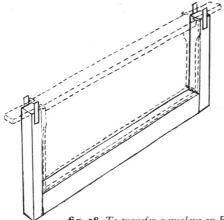

fig. 28 *To transfer a nucleus on British Standard frames into a Modified Commercial brood chamber an adaptor can be used, constructed as shown here. The metal strips are bent around the lugs.*

of year when bees may be expected to draw out foundation. The beekeeper should be in no hurry to work out the adapted frames but can do so over a couple of seasons.

Some beekeepers keep their bees on 'brood and a half', that is they give the queen the run of a National brood chamber and one super. In this case the first super to be put on the nucleus will be an extension of the brood chamber and therefore should have frames spaced as in the brood chamber. The queen excluder is accordingly put below the second super. I consider brood-and-a-half a messy system, and not to be recommended.

The new beekeeper with a full brood chamber and a super above now owns a full colony. He should keep putting on supers as these are required until the end of the season—usually the end of July to the end of August, depending upon his particular area. If he asks a few of the local beekeepers he will get a good idea of what to expect. With any luck he should not have to deal with swarming in his first year, if the nucleus he obtains has a new young queen, but mistakes are often made by the beginner and he should keep his eyes open for queen cells. A beekeeper with a second hive ready is prepared for any swarm and this can be dealt with as described in Chapter 7. Similarly, if queen cells are seen before they have swarmed an artificial swarm could be produced, as described on page 138. Either of these methods will produce a second colony in the spare hive and will prevent any repetition of queen-cell production for the rest of the season.

5 The year's work

We must now consider the management of full-sized colonies: in other words, the management of bees in any apiary and in any quantity. The problems affecting the individual colonies will be the same no matter how many you have, and the answers will vary in a quantitative, not qualitative way. I want first to deal with the year's work in the apiary in broad terms, leaving the details of manipulation to the next chapter, in order to show beekeeping as an on-going process in which the beekeeper is trying to make his colonies productive units from which he can take a crop.

My training and inclination is towards getting each colony to produce about the same amount of honey, if possible. This is the *intensive* method of keeping bees rather than the *extensive*, where more colonies are kept than can be adequately controlled and a planned percentage of unproductive colonies is accepted. Carried to an extreme the extensive method produces the 'leave alone' beekeeper who never looks at the brood chamber of his colonies but merely puts on supers and takes any crop he is lucky enough to find at the end of the year. This is antisocial behaviour on the part of the beekeeper, as he runs the risk of retaining disease in his colonies for unnecessarily long periods, and makes his colonies a possible source of danger to his fellow beekeepers, as well as spraying the neighbourhood with stray swarms which are a nuisance to everyone. It is impossible to call this 'owner of bees' a beekeeper. Any beekeeper should look after his stock and ensure it is not a source of trouble to anyone else, beekeeper or non-beekeeper. The methods set out below therefore entail some work, although this will be kept to a minimum, and can be reduced somewhat as the practitioner becomes more experienced.

The beekeeping year starts at the end of August and the beginning of September when the honey has been taken off and the colonies are being got ready for winter. The aim is to see that every colony has a

Heavy work. When the weight is honey no one minds.

good start the following spring by going into winter in optimum condition. What have we got to do? What is the optimum condition for a colony? The colony should have a young queen and plenty of bees. Stores should be sufficient to last them until the weather warms up and the spring flowers arrive. The colony should be free of disease, and protected from predators and pests. The bees must be in sound, waterproof hives so that they are dry, and preferably on stands with good air circulation around them, situated in a dry, warm, unexposed apiary. If they are not, then the beekeeper should endeavour to improve matters as hives are put to the test, particularly in winter.

There is considerable advantage in having young queens in the colonies for wintering, partly because they are less likely to die or to become drone breeders but mainly because the younger they are, the later in the season they tend to keep an active broodnest, which means that these freshly-emerged workers do not have to live as long under winter conditions. It is probably for these reasons that the old skep beekeepers used to say that the first *cast*, that is the *second* swarm from a colony which would contain a young, just emerged, queen and a lot of young bees, was likely to be a very good colony the following year. It is certainly true in my experience that a young queen with a lot of young bees in a freshly-established nucleus often winters better than fully-developed and long-standing colonies. Therefore I would keep queens for only two seasons, getting rid of them at the end of this period, and requeening (see p. 157) in about the first week in September before feeding down for winter. You will have to adjust this timing to suit your own area. Some beekeepers keep queens three years quite satisfactorily, but for the beginner I would advise keeping queens for two years only, until sufficient experience has been gathered to find out whether they will last without failing in the middle of the third season.

The next question will be, 'Where do replacement queens come from?' You can of course buy them, but as I believe that the provision of the next generation is part of beekeeping, I hope you will try to produce them for yourselves. Details of how to do so are given in Chapter 8.

Removal of two-year-old queens should take place in about the last week of August or first week of September. You will usually find that two-year-old queens will have given up laying at the end of August and there will be no brood present, but in some colonies where there is an old queen, two, three, or more frames of brood will greet your eye. This is where supersedure has taken place. Careful searching will reveal a new young queen and often the mother will be there as well, the two of them laying away, usually on the same comb. This is quite common in colonies. If you come across the condition of an active broodnest where you know there is an old queen and you find her, do not jump to the conclusion that it is she who is doing the laying and replace her with a new queen. If you do this, your new queen will be killed because you can be certain there is a new young queen doing the laying. Go on and find her, to satisfy your mind, and leave her to continue her good work.

Colonies which for some reason or other are small in number of workers should be united in autumn and given a new queen. The cause of small colonies may be either poor queens, too much swarming, disease or accident. Uniting can be carried out using the paper method (see page 163), but care should be taken if disease is the cause of low numbers. Any beekeeper suspecting disease should read Chapter 9 carefully and take appropriate action.

Having introduced a new queen successfully, if this is part of the necessary preparations for wintering, the next job at about the end of the first week in September is feeding the colonies for the winter with sugar syrup, as described on pages 130–33. The aim here is to have in the hives sufficient stores to last the bees right through to the next active season—at least until April—without having to try to feed either syrup or candy to them at all during the winter period. Feeding candy is detrimental to the bees as they have to fetch in water to liquify it, or use the output of the salivary glands which exhausts the colony, and in any case is a nuisance to the beekeeper.

How much syrup should each colony be given? This is a question always asked by the beginners and unfortunately there is no single quantity which can be stated in reply, because it will depend upon the amount of honey already in the brood chamber. The beginner must look through his colonies and assess the amount of stores by eye. He can use the knowledge that a British Standard brood frame when full on both sides will hold about 5 lb. of honey; a piece of brood comb 3×4 inches full of sealed honey will hold about $\frac{1}{2}$ lb. honey; an inch depth of sealed honey right across both sides of a brood frame will hold about $\frac{2}{3}$ lb. honey. Using these quantities the beginner can go through his colonies and calculate roughly how much honey is in the brood chamber. As he assesses the colonies in this way he should lift the hives and so learn to evaluate the amount of stores by weight from outside. Experience of judging the amount of food by 'hefting' the hive comes remarkably quickly.

How much honey should each colony have by the time feeding is ended? This will vary with the strain of bee, and is correlated to the size of brood chamber the queen uses in the summer. My recommendation is to aim at 40–45 lb. of stores for a colony in which the queen does not need more than a National brood chamber in the summer, and 50–60 lb. for bees whose queen fills a National brood chamber and super, or one of the larger brood chambers. Therefore having looked through such a chamber and assessed that the colony has, say, 30 lb. of honey, an extra 20–30 lb. of honey-equivalent in syrup will be needed. Two pounds of sugar is the equivalent of $2\frac{1}{2}$ lb. of honey, and so the requirement will be provided by 16–24 lb. of sugar made up into syrup. Get this into the colony as quickly as possible, using a large feeder (see page 130).

Feeding is best done in the evening so that darkness will help quell the rushing about of the bees which always occurs when feeding is taking place. It will also reduce the chances of starting robbing, as will reducing entrances in size by putting in the entrance block. If nuclei are present in the apiary it is often advisable to reduce their entrance to about a couple of inches. It is always advisable to feed all colonies in the apiary at the same time, using several feeders, as feeding puts all bees on the alert and multiple feeding does not present such an invitation to robbers from unfed hives. Should there be signs of robbing—bees fighting and flying in an erratic fashion, trying to get in without contacting the occupants of the hive—then entrances can be cut down to a single beeway (about $\frac{1}{4}$ inch) so that the guards can emulate Horatio in their gateway.

Bees usually manage to store enough pollen for the winter, but if a beekeeper with several colonies finds that a particular colony is short of stored pollen he may be able to take a comb containing a lot of pollen from another colony to give to the colony which is short. I would myself rather try to ensure that an early and late pollen crop is available to the bees, and the beekeeper will find it well worth his while to plant flowers in his garden which are good sources of pollen, such as hazel and willow for spring and Michaelmas daisies and ivy for autumn. Other useful plants are mentioned in Chapter 10, on forage flora.

Once feeding has been finished the hives should be made fast against predators and pests. Mice are usually a great problem and will get into any hive which is not fitted with a mouse guard. Some six years ago I saw a colony in October without a mouse guard, and noticed that pieces of comb were being pushed out of the entrance. I lifted up the brood chamber and looked underneath, and there were five long grey tails hanging down. I put the brood chamber down again and gave it a kick. Five jet propelled mice came flying out of the entrance with several very angry bees attached to each. It is quite extraordinary that the bees did not attack them until I did something which released the attacking impulse. However, it is better to put mouse guards on before the mice start to come indoors for the winter.

Mice are excluded from a hive by making the entrance a round hole not more than $\frac{3}{8}$ inch in diameter or a horizontal slot no higher than $\frac{5}{16}$ inch. The mouse's skull is wider than it is high so that although it cannot get through a $\frac{3}{8}$ inch round hole it can get through a $\frac{3}{8}$ inch high slot. Guards of many sorts provide entrances of this sort: strips of metal with $\frac{3}{8}$ inch round holes can be purchased from the equipment suppliers; pieces of perforated zinc can have a slot cut into them; the

Woodpecker damage.

wooden entrance blocks can be used providing the doorways are the right height, and do-it-yourself geniuses can think of many other ways. Do not, however, use a queen excluder. Although this is quite efficient as a mouse excluder, it has the disadvantage of knocking loads of pollen from the legs of the bees as they enter after foraging, at a time when the value of fresh pollen is at a premium.

Another winter pest that you may have to deal with is the green woodpecker. Woodpeckers learn that they can find a good meal in a beehive much in the way that bluetits learn to open milk bottles for the cream. You may keep bees in an apiary for years with lots of green woodpeckers about without any damage and then suddenly they learn the trick and through the hive wall they go, leaving behind a dead colony and several 3 inch holes. Whether all the damage is done by the woodpeckers or whether rats finish the job off I am not sure, but I have seen brood chambers in which the frames have been turned into a pile of wooden splinters, no piece being larger than a match. Covering the hive with wire netting or fish netting before the first frosts is the usual remedy.

Colonies which have been put down to winter as recommended above should now be safe to be left entirely alone until the following April when the active beekeeping season starts once more.

Before leaving the subject of wintering, however, I would like to mention ventilation. Many articles have been written on this subject without any definite conclusion being reached. I feel that there is no real indication as to whether a lot, a little, or none is best for the bees, either from my own experience or that of other beekeepers with whom I have discussed the subject. I would tend to use more ventilation in relatively warm, damp areas than in the cold, dry regions. Ventilation can be effected by raising the crown board upon $\frac{1}{4}$ inch blocks at the corners. This allows the wet air to flow out at the sides without creating a draught through the cluster. Try ventilating some colonies in winter, and see if it suits your bees over the years.

Before the active season starts again the beekeeper should do any repairs to spare hives and equipment that are necessary and prepare for any increase by making or buying the necessary hives. A couple of brood frames should be prepared ready with foundation for each colony. These will be needed in due course to replace broken frames of comb and those with too high a percentage of drone cells in their make-up. And then, most important of all, the beekeeper should *plan* his beekeeping for the coming seasons. Planning cannot of course be in detail, as no two seasons are exactly alike nor two colonies exactly the same and variations in types of forage available make many decisions necessary as one goes along. The successful beekeeper must be a great opportunist, to take advantage of helpful circumstances as they arise. However, all these decisions should be made within a general plan which is drawn up before the season begins. You must decide what you are going to do about poor colonies, how you are going to provide replacement queens, how you will organize increases in numbers of colonies, what you will do with colonies which may try to swarm, and whether you have you enough supers for a good year.

If a colony has for some unavoidable reason been wintered with less stores than the recommended amount, it is a good idea to heft or lift the hive a couple of times in March to make sure that it is not getting too light in weight. If a colony seems too light it should be fed, using a contact feeder made of plastic with a patch of wire gauze in the lid (see page 131). These are excellent and are sold by some of the equipment firms. In cold weather winter bees will only feed from contact feeders, and will not enter aluminium rapid feeders or Miller feeders.

Once April arrives the year's work begins in earnest and examination should start. It is the beekeeper's job to assist the broad pattern of the annual colony cycle as shown in the graph on page 49, and to make this conform to his requirements so that the maximum crop of honey is available to him at the end of the year. In the early part of the season he has to assist the colony in building up. During the peak period he has to defeat the desire of the colony to swarm, should it try. He has to ensure that the colony has sufficient room to accommodate its rising population and to store all the honey it can forage, and that after he has removed the honey crop it has enough stores to survive the winter.

The only way to cover all this is to examine the colonies regularly and to assist those that need help. Normally these examinations need not take more than about seven minutes per colony for a man working on his own, or three minutes for two men working together. The beginner will take longer of course because there is so much he wants to look at which is not really necessary in the routine examination of the colony. He needs to do this to learn the craft, and even the experienced beekeeper may sometimes want to look carefully at some aspect of the colony which is outside the routine inspection. These occasions when the colony is left open for a longer period should only occur at times when conditions are ideal, warm and sunny, and when the beekeeper has plenty of time.

When you go to your bees in April and open up the hive as described in Chapter 6, what are you going to look for as you work through the colony? You have to answer five questions each time you look at a colony on routine inspection. These are:

1   Has the colony sufficient room?

2   Is the queen present and laying the expected quantity of eggs?

3 a (early in season) Is the colony building up in size as fast as other colonies in the apiary?
   b (mid season) Are there any queen cells present in the colony?

4   Are there any signs of disease or abnormality?

5   Has the colony sufficient stores to last until the next inspection?

The answers to these questions will give the beekeeper all the information he requires to work the colonies. If the answers to questions 1, 2, 3a, and 5 are affirmative, and to questions 3b and 4 are negative, the colony needs no special work. If any answer is the opposite to the above some action is required. Detailed analysis of these questions and the methods of dealing with problems is given in Chapter 6. Here we will only look at them in the broad principle.

The first and last questions are very easily answered and the solving of associated problems very quickly accomplished. If more room is needed, a super should be given. If colonies are light and need feeding, a gallon of syrup (8 lb. sugar in water) should be given in a rapid feeder.

Question 4 on disease is in a different category because the recognition of disease depends upon both knowledge and experience, and should be answered in the light of the information given in Chapter 9. My advice to the beginner is to study the diseases in Chapter 9 and any modern books on the subject he can obtain. When looking at his colonies he should note the normal appearance of the various aspects of the bee society which change when disease occurs. For instance he should look at the shape and colour of the capping on brood cells, the colour and position of healthy larvae in their cells, and the appearance of bees at the hive entrance. By getting to know the look and shape of the normal individuals in the colony he will spot anything which is abnormal. Having noticed something unusual he should try to find out the cause and confirm a diagnosis with a competent beekeeper. Do not be afraid of doing this. Ministry Foulbrood Officers, County Beekeeping Lecturers and any really experienced beekeeper will rather be asked about something which turns out to be nothing than to have beginners nursing worries and possibly harbouring disease.

We are then left with questions 2 and 3a, which contain the real crux of colony development. If you visit a number of apiaries in spring you will find the same general picture. One or two colonies will be really thriving, with a large force of bees flying and about seven frames of brood. The large majority will be doing moderately well, with five or six frames of brood, whilst some will be poor, with only three or four or even fewer frames of brood and a meagre force of flying bees. The proportion of this 'unproductive tail' will depend upon the season and the quality of the beekeeping. Poor seasons and poor beekeeping will increase the size of this tail; good seasons and good beekeeping will reduce it, but will not get rid of it entirely. There is always some variation in the size of the colonies and honey production by and large is in proportion to the size of colony in an apiary where bees are all of one strain. (This does not hold good, however, if comparisons are between colonies of different strains.) Comparison is a great aid in beekeeping practice. If one colony can do well why cannot all the rest be just as good? What is holding them back? Does the problem lie with the presence of a poor queen in the colony or something else? Questions 2 and 3 are intended to set the beekeeper on to the right track to assist the colony.

To answer question 2 there is little need to find the queen. The presence of eggs in quantity means that she was there within the last three days, or all the eggs would have hatched. It is very unlikely that she has died in the meantime. In the early season eggs are sufficient evidence that she is there. Is she laying the expected number of eggs? What number would you expect her to be laying? If you have more than one hive, then assessment can be made by comparing her output with that of the queens of the best colonies in the apiary. If she is not doing as well as the best she may lack adult bees to look after the brood and so is working below her full potential. Conversely, she may have plenty of bees, who should be looking after the brood and are clustering instead on empty combs because she is incapable of increasing her egg lay rate.

The answer to question 2 affects question 3a. Is the area covered by brood increasing and extending partially to cover more combs, and is the area of brood well covered with bees, with other bees working on empty combs and ones filled with honey? The brood should show evidence of the queen's laying in concentric circles with larvae of the same age in adjacent cells, not of mixed ages occurring in cells next to one another. Sealed brood—larvae undergoing metamorphosis into adult bees—should be tightly massed in an area with very few unsealed cells. Scattered age of brood indicates a poor queen, perhaps of reduced fertility or partially paralysed. Such a problem can be solved by requeening the colony with a young laying queen. This is the only answer: if you let the old one carry on, the colony will probably be unproductive for the season. The bees themselves will solve the problem by supersedure, but too late to keep the colony productive in that year—their only concern is to keep the colony alive to carry on the species. If the beekeeper forces a colony of this sort to produce a queen for itself by killing the old queen the new queen he will get will be small and very poor in quality. (See Chapter 8 on queen rearing.)

Disease such as nosema is another reason why colonies fail to build up at an active rate. If the queen appears adequate, check for disease by referring to Chapter 9. In my experience it is very rare to find colonies failing to develop properly which cannot be assigned to either the 'poor queen' or nosemic categories, although recent work by Dr L. Bailey suggests that there are quite a few virus infections being researched which may give us some added understanding of wintering problems in the future.

Spreading the brood

The majority of colonies in spring are small to average in size, and we may well wish to boost these up to being as good as the best. The distinctive feature of such colonies is that they are building up, albeit slowly in some cases, but sufficiently for this to be visible between examinations at weekly intervals. We can decide if their queens are in good condition and capable of increasing laying if given the chance, and we can check that the colonies, as far as we are aware, are not being held back by any disease.

> **Overleaf** *A beautiful comb of brood which shows a methodically laying queen. At the top left the cells are sealed. The larvae are progressively younger across the frame, and recently laid eggs are seen bottom right.*

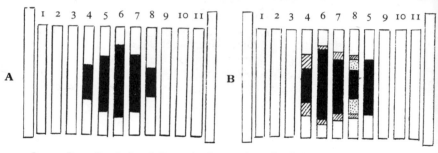

fig. 29 *Spreading the brood. By moving frame 5, immediately induced expansion is shown on frame 8, with natural expansion hatched.*

These average colonies can be induced to build up more rapidly by a process known as 'spreading the brood'. This increases the rate at which the queen will lay above that which would occur if they were left to themselves. Spreading the brood must be practised with understanding. The process is illustrated as follows by the colony shown in fig. 29A. This colony is building up slowly, with no hindrances, and there are plenty of stores. The dark portions show the size of the brood nest which is on combs 4 to 8. These are well covered with bees and there are plenty of bees on the stores outside the broodnest. The brood is examined and a medium-sized comb of sealed brood near emergence is selected. In this case comb 5 is chosen and moved to the position shown in B, next to comb 8. The result is that the queen will lay up the pieces shaded as 'induced expansion'. The natural expansion of build up will continue as shown. Some of the expansion on comb 4 may also be induced but it is difficult to assess. Notice that the number of combs in the broodnest has not been increased, but that the rearrangement of combs has caused the broodnest *area* to be increased. This increased area of broodnest was warmed up by the heat from the brood in the comb which was moved ('the spread'), and it is probably the redistribution of heat which causes the expansion, and at very little extra cost to the nurse bees who control the temperature of the broodnest.

The concept of brood spreading is carried a stage further in fig. 30. The colony here has six combs with brood on them and has started against one of the hive walls. The brood is examined and comb 3 is selected as a good comb of emerging brood, and moved to the outside against comb 6. The resulting expansion on the edges of combs 5 and 6 is shown, and with a reasonably lively queen expansion will continue out on to another comb—No. 7. If we now renumber the combs in diagram **b** in linear order, we have the same broodnest illustrated in **c**. Again, selection of brood is made as normal inspections are carried out, and this time comb 4 is the er̃ ̃ing comb of brood, which is moved to a new position next to comb 7. The induced expansion occurs mainly on comb 7 but also on combs 5 and 6 and natural expansion is shown on combs I and 2. Increase is only in an area within the broodnest which is not extended on to another comb. Because the broodnest starts from one wall of the hive, any extension on to new combs must occur on the right hand side of the broodnest in fig. 30a, whereas in fig. 29A extension can occur on either or both sides. For this reason the colony in fig. 30 can be assessed for expansion much more quickly and accurately than can a colony set up as in fig. 29. It is therefore my practice to move the broodnests in each hive over against one of the walls, and in such a way that when working the colonies the brood is furthest away from the operator. A single-handed beekeeper with

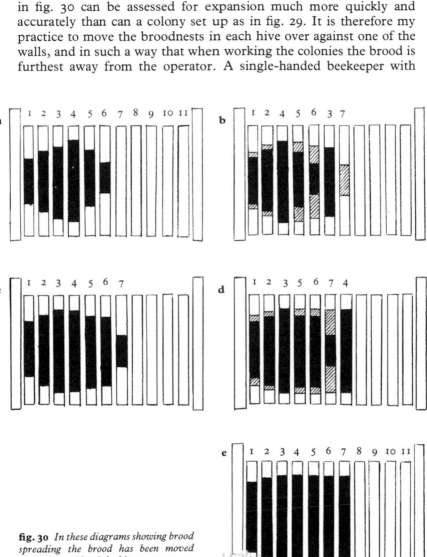

fig. 30 *In these diagrams showing brood spreading the brood has been moved against one side of the hive to encourage expansion in one direction only.*

hives on double stands would therefore move the broodnests towards the centre of the stand. If fig. 29 is taken as an example of a pair of hives on a stand side by side, then in hive A combs 9, 10 and 11 would be removed, the rest of the combs moved over to bring comb 8 against the wall, and the removed combs put back in again so the comb order would now be 9, 10, 11, 1, 2, 3, etc. Colony B would be treated in the opposite way, removing combs 1, 2 and 3 and bringing 4, etc. over against the left hand wall. On approaching the colonies you would know that all the empty combs or combs full of stores will be nearest on opening up, with the broodnest on the far side. As the crownboard is lifted it is thus possible to get a good idea of the size of the colony immediately by the coverage of bees on the top bars. Furthermore, the first frame of brood will indicate the expansion which has occurred since the last inspection.

A possible danger can be avoided by studying fig. 30**e**. This is exactly the same broodnest as in **d**. It is now almost squared up, having lost the spherical shape of the earlier broodnest in **a**. There would therefore be no advantage in moving a comb of brood for the purpose of increasing the queen's laying because there are no small combs of brood. The colony has not expanded on to another comb, and the queen cannot be forced into laying extra eggs. The natural extension of the brood-spreading method would be to introduce, say, comb 4 in between combs 8 and 9, thus enclosing the completely broodless comb 8 within the broodnest area. This may be disastrous. This comb has not been prepared for the queen to lay in and instead of the bees adopting it as part of the broodnest it may form a barrier which the queen will not cross. Further, if the colony is smaller than the one shown here, not only may the empty comb form a barrier but the bees may not be able to maintain the temperature of the spread comb if the weather should turn chilly at night. Dead brood would be the result. Appropriately enough the beekeeper would call it 'chilled brood'. The moral of this is therefore never spread brood *over* an empty comb.

I would however still spread the brood in colony **d** but for other reasons than that of increasing the laying rate of the queen. Now let us examine what actually happens when brood is spread. Comb 4 in **c** contains sealed brood, some of which is emerging as adult young worker bees, and has been placed between combs 7 and 8. The queen will fairly rapidly lay up the broodless area between combs 6 and 4, which is shown hatched in **d**. She will also probably be expanding the whole broodnest by her normal increase in laying rate. The additional eggs are produced by the better distribution of heat in the broodnest. But what happens to comb 4? The bees go on emerging from it as the days pass. If the hive is a Modified Commercial hive, the full frame will contain about 3,500 cells on one side or 7,000 on both sides together. The comb is, however, only about $\frac{3}{4}$ full of brood, thus approximately

5,500 pupae are present, some only just emerging. The queen at this part of the season—late April or early May judging by the total size of the broodnest—will be laying about a thousand eggs a day. Comb 4 will therefore have taken about $5\frac{1}{2}$ days for the queen to lay up, and it will take the same period for all the bees to emerge, as the length of brood cycle is very constant. As the bees emerge they will clean up the cells and the queen will start relaying the comb. If we allow a day for cleaning up and relaying then the spread comb will have completely emerged and been relaid by $6\frac{1}{2}$ days from the day the comb was moved. If you are looking at your bees every 7 days, you will only have to look at combs 4 and 7 in colony **d** to see most of the work done by the queen since you last looked at them. If comb 4 has been relaid then she is increasing her laying rate because she will also have relaid any other areas of the brood which have emerged. You will also be able to judge the quality of her laying by the evenness of the ages of larvae in adjacent cells, and the percentage dying by the number of empty cells within the brood area, or cells with eggs surrounded by blocks of larvae.

If you look at your bees once a week, examination of two combs will answer the questions about the queen and the colony's build-up. Quantity of stores is obvious and the examination can be concluded by 'spreading' the brood by moving one comb for the next time and shutting down within about three minutes of first opening. The bees will hardly realize you have been, and foraging will continue unabated or very little reduced.

Spreading the brood should only be done on colonies which are building up, however slowly, where there is a foraging force, and when there are bees working on some of the empty combs. Very poor colonies should be dealt with as indicated on page 127, but by providing good young queens and dealing with nosema by feeding Fumidil in the autumn when it is required such colonies should be almost non-existent in your apiary.

The process of brood spreading is continued by examination until the queen is laying a broodnest which extends over the whole width of the brood chamber. Before the bees spread out on to the last combs a super is put on to give extra room, and further supers are added as needed. If the queen does not re-lay the spread up to mid season, she is failing and should be replaced, stressing the need to have young queens available at all times. Once the colony has a full-sized broodnest it is 'built up'. This state should be reached by the middle to end of May in southern Britain, but this will vary for different districts and the beginner should enquire of beekeepers in his area when they expect colonies to fill the brood chamber. As the season progresses the interest in the building up of colonies and the solving of the problems that are holding them back gives way to the problem of preventing the loss of swarms. Thus, when examining colonies, question 3b replaces 3a: 'Are there any queen cells present in the colony?'

This introduces us to two new management techniques regarding swarming, or rather its prevention: firstly, the clipping of the queen's wings in the early part of the season, and secondly, rigorously timed examinations to ensure that the beekeeper does not miss queen cells, once the colony has built up to a full-sized broodnest. Let us look at these techniques.

Clipping the queen's wings means cutting off enough to prevent her flying. It is of enormous help to the beekeeper, giving him extra time to play with when colonies are making queen cells and thinking of swarming. This time is gained by the fact that clipped queens swarm when the first virgin queen is ready to emerge. It is extremely rare for them to swarm out before this time. Unclipped queens, on the other hand, will often swarm out when the first queen cell is sealed and often will go earlier than this if disturbed by manipulations. Clipped queens, if they are allowed to swarm out, fall to the ground and are lost by their followers. The bees often hang up nearby for twenty minutes or so and then return home. You may therefore lose the queen but you will not lose your bees (and it is they who gather the honey) until the first virgin queen is on the wing and can lead the swarm away. Using the delay produced by having clipped queens we can tie this in with the period of development of a new queen to establish a maximum period between inspections which will ensure that no swarm gets away, providing always, of course, that we do not slip up when using any of the techniques required.

The queen cell is sealed 8 days after the egg is laid. The virgin queen emerges 8 days later, on the sixteenth day, and may fly the next day. Therefore if a colony has a clipped queen and is not making queen cells at one examination it will not be able to get a swarm away for 17 days—indeed it is unlikely even to lose its clipped queen before the sixteenth day. In addition, if a colony is making queen cells at one inspection, and these are killed by the beekeeper, then emergency cells will probably be started from worker larvae. If a 4 day-old larva is chosen (and this is probably the oldest they would choose), then this larva, being 7 days old (3 days in the egg and 4 days as a larva), should emerge as a queen after 9 more days (16 days minus 7 days) and can be on the wing on the tenth day after inspection. The fact that the bees will usually select a larva at least 2 days younger means that usually the virgin will not emerge for at least 11 days.

If colonies are being examined every 7 days for convenience during the building-up period, we can cut down the number of inspections needed once the colonies have built up by using clipped queens, because any colony not making queen cells one week cannot get away with a swarm for over 14 days. We can therefore miss the next inspection on any colony which is *not* making queen cells. I would always try to make as few examinations as possible, not only to save time but also to save disturbance to the bees.

If, on the other hand, it is convenient to inspect colonies at any time, and you wish to stretch the periods between examinations of all colonies to the maximum it is obvious, from the figures above, that this maximum is 10 days. Thus using clipped queens and examining every 10 days will prevent the loss of swarms entirely, providing that no queen cell ever slips by unobserved. In fact, in inclement weather it is possible to leave inspections for 11 or 12 days and get by without having a swarm. However, the colony which swarms and from which the swarm has been lost usually produces little or no crop that year, and so inspections must be made rigorously to time.

Swarming can be held back by giving plenty of room ahead of requirements by adding supers, but there comes a time when colonies in the apiary will be found with queen cells. Some method of swarm control must then be used or the yield from these colonies will be drastically reduced. We know from work that has been done in America, and from experience, that not all colonies will swarm even if they produce queen cells. Unfortunately as yet no one has been able to give us a method to distinguish between those that will and those that will not. Therefore until such knowledge is available, and capable of practical application in the apiary, we have to treat all the colonies as though they are going to swarm.

Two methods of swarm control are given in detail in Chapter 7. These are requeening the colony with a young queen, and the artificial swarm method. Because I try to keep my beekeeping as simple as possible and to have everything under control, my preference is for the first system. This, of course, does depend upon having young mated laying queens at hand all the time. Their production will be dealt with in Chapter 8. The beekeeper should make up his mind what he is going to do about swarming so that when colonies making cells are found in the apiary he automatically goes into the selected routine. The beginner should refrain from mixing various systems of swarm control as this requires considerable experience. In his first years he should stick to one method until he feels that he has a basic understanding of the honeybee and its colony. Once he has reached this state experimentation with various methods will give him a lot of enjoyment and still further increase his understanding of the bee.

The beekeeper must look for queen cells at each examination. Right through the active season from mid April onwards colonies will make small queen cell cups (see overleaf). They can occur anywhere, but are usually on the bottom and sides of the brood combs. When colonies are being examined you should look into these cups to see if they have eggs or young larvae in them. You will notice they go through a

In one of these incipient queen cups is a larva and in the other is an egg, but it is sometimes difficult to see without tearing the side out of the cell.

development from a cup with a dull matt internal surface to a polished stage, ready for the queen to lay in, to the stage where they contain eggs or larvae. I would not treat them as queen cells until they contain a larva because often eggs in these cells are not allowed to develop further by the workers—I suspect that they eat the eggs until conditions are appropriate for queen cell production.

Look at the cups—the quickest method is to tear the side out of them—and if one has a larva in it then your chosen swarm prevention method must be adopted. If you have decided to use the artificial swarm method, the colony should immediately be split. The brood and queen cells with the house bees (in this case those that cannot fly) are put in one brood chamber, the queen on one frame of brood in a another brood chamber full of drawn comb, which is left on the old site and given the supers. All the flying bees will return to her as she is on the old site and the colony should carry on storing honey if there is a nectar flow. This is one of the methods based upon splitting the colony which has been used for many years to prevent swarming. It attempts to do the two things necessary to deal with a colony about to swarm: to prevent it from swarming out and to requeen it by the end of the process. It is quite successful although, as with all programmes which deal with living things, it sometimes fails.

Because I prefer to keep the colonies together in one piece, on finding queen cells with larvae I would remove all of them and leave the colony for a further examination period. If they are making queen cells at the next inspection I would destroy these again. It will be found that many colonies will give up queen cell production after these have been destroyed once or twice. In fact about 25 per cent of colonies who produce queen cells can be dissuaded by this method. The rest will definitely have made up their minds and so if they are making cells again at the next inspection, I would *kill the queen and queen cells*, leaving the colony queenless for a week at least, and then requeen using the nucleus method. The advantage of this method is that it is more controlled: only those colonies which are really intent on swarming have to be treated. The only drawback to this method is that you must have young laying queens available to requeen with.

Colonies in an apiary tend to come to a peak at about the same time and so the swarming peaks tend to coincide as well: usually most of the colonies that are going to produce queen cells do so within about a fortnight of one another. Once the swarming period is dealt with in an apiary it is usually over for the year. Sometimes a very small percentage produce cells late in July but the main period will be in June, and once over no further serious trouble is to be expected.

If you have a number of apiaries fairly widely separated it is noticeable that the swarming peak comes at different times in different districts. This is really only to be expected in an animal as tightly tied to its environment as the bee. Most of the swarming occurs when there is very little honey flow and the starting of a large honey flow seems to bring swarming problems to an end much more rapidly than in a long period of dearth.

Having got over the swarming period there is a fairly quiet time in the apiary for the month of July and the beginning of August. Supers are added as the colonies require them, that is *before* the bees are covering all the frames they have available. Always keep ahead of the bees. Remember nectar takes up two to three times as much room as finished honey so that extra room is required for storage while the nectar is being processed. The number of supers required will depend upon the area and the weather prevailing at times when forage plants are in bloom. It is advisable to have three supers available per colony as an average, but in a good season this will not be enough—I have yet to meet the beekeeper who has all the supers he requires in a good year. Where oilseed rape is a general forage crop for the bees supers have to be removed, the honey extracted, and the supers replaced before the honey granulates in the comb (see Chapter 11). Similar problems occur with other crops such as mustard and kale, both related to rape, and I believe the honey from raspberry also suffers from rapid granulation. Otherwise all honey is usually left on the hive to the end of the season.

The beekeeper will have to enquire in his own area regarding the end of the local honey season. Very considerable local variation in the date occurs. Some areas of the south of England are finished completely by the end of July and in others considerable flows occur up to the second week in September in some years. Ivy produces a very late flow in several areas, sufficient to supply a high proportion of winter stores. Local beekeepers will be able to tell you what is usual but it is as well to remember that a local farmer can completely change the pattern by altering his cropping. For this reason it should not be assumed that amount and timing of honey crops will always remain the same.

When you know you are within about a fortnight of the usual end of the honey flow, refrain from adding supers to those colonies which still have some spaces to fill in supers already on the hive. By this means it is possible to cut down the number of partially filled supers for extraction, which reduces the amount of handling. Also, it will be found that full frames are more easily decapped than partially filled ones.

Once the flow is over the honey will be removed for extracting, and the method of removing the honey supers and clearing the bees away from them is described in detail in Chapter 11. Before removing the honey you should, however, make quite certain it is ready. Supers containing only sealed honey can be removed as soon as you wish. The problem is often the last super put on the hive, which may have a considerable amount of uncapped, or unsealed, honey in its cells. It used to be recommended that unsealed honey be extracted first and later fed back to the bees. This is nonsense, based on the fallacious idea that unsealed honey was below the density required of finished honey. In fact the honey will be fully up to gravity a couple of days after the flow has ceased: it will be unsealed because the cells are only partially filled, and the ever-hopeful bees are waiting for more nectar to come in to finish filling them. To test whether the honey is fit to take off, take a frame with plenty of unsealed honey and give it a good heavy shake towards the top bars of the super. If any drops of honey fly out, leave the super for a few more days and repeat. When the honey is of finished specific gravity no drops will fly out. You must only do this with well-wired frames and well-attached comb or the whole comb may fly out when you shake it.

Having removed the honey we are back to the beginning of the beekeeping year. We have come full circle, and begin again with wintering down, as described on page 97.

Recording

If you keep simple records you can check on what was happening in the colony when you last saw it, and if you keep them for long enough you can see where you have made mistakes in colony handling season by season, and avoid them in future.

From the practical point of view, it helps if records can be taken in at a glance. They may be kept as a separate card left on each hive and collected up at the end of the year, or each as a separate leaf in an apiary book. One form is illustrated below. Top left shows the origin of the queen (A); the figure 6 is the last digit of the year she was produced, and the cross denotes that she is clipped. On the other side is the colony number—essential if you are collecting the cards at the end of the year. The five columns after the date column refer to the five basic questions mentioned earlier and will act as a reminder when you first start keeping bees of the information you must seek when examining. It is quite easy to go through a colony and be so interested in its development that to check for the amount of stores is forgotten. A note on the colony growth can be kept in the remarks column by entering the number of frames containing brood. Any syrup given can be noted in gallons in the extreme right hand column. When the colony starts to produce queen cells these are noted in column 3 by means of a simple code: 1 = eggs in queen cups, 2 = larvae in queen cells, 3 = sealed queen cells. In this way a glance at the record card gives all the information required and time taken in writing up records is minimal.

Queen A6x						Colony No. 3	
date	*1*	*2*	*3*	*4*	*5*	*remarks*	*feed*
15/4	/	/	?	/	/	5 fr brood	
25/4	/	/	/	/	/	6 fr brood	1
15/6	/	/	12	/	/	destroyed Q cells	
25/6	/	/	123	/	/	destroyed Q cells	
5/7	/	/	123	/	/	killed Q, requeen nuc B6x	
						Crop 50 lb.	

specimen record card

6 Handling the bees

Before opening a colony for examination, collect together *all* the things you are likely to need for the manipulation, such as hive tool, queen excluders, supers, etc. Light the smoker. The fuel used may be any solid material that lights easily and burns, or rather smoulders, producing plenty of cool smoke for a reasonably long time, so that refuelling is not a constant need. The usual fuels are corrugated paper, sacking, dried grass and rotten wood. Quite a lot of corrugated paper appears to be fireproof these days, but if it will smoulder easily it should be rolled into a cylinder to fit the firebox of the smoker. Sacking lasts longer but care should be taken that it has not been used to contain anything poisonous to the bee, such as dressed grain. I prefer grass which has been cut with a rotary grass cutter and left to dry. This can be picked up and stored in a sack, and two of these will last me a season. It has the advantage that it burns nicely and with little residue in the smoker, and if the smoke does blow back into your face when lighting it is not as vicious as sacking or paper smoke. I usually start the smoker by setting light to a small ball of newspaper. Ensure the smoker is going well; put on your veil and gloves. Make your way to the colony quietly. Do not cause disturbance by stomping around the hives, nor drop your extra equipment on to the ground: put it down quietly, as bees readily detect vibration.

Gently smoke the entrance of the hive. Do not puff smoke in until the bees come out crying; let the smoke drift in. The smell of smoke causes the bees to fill themselves up with honey from the honey store, and this renders them much more amenable to handling. A full bee, like a well-fed human, is much less likely to want to start a fight. It takes about two minutes for the bees to fill up and for the full effect of the smoking to be obtained. Beginners are therefore advised to take things steadily and to wait this amount of time, giving the bees a reminder in smoke two or three times. The beekeeper, as he becomes

Handling a large colony with assurance comes with practice.

more experienced and confident in his handling, will find that smoking at the entrance can be cut out entirely, smoke being applied under the crown board as this is removed. This saves time, and it is usually just as effective, but the beekeeper should learn to keep one eye on the entrance because every now and then a colony may start to flood out from here and a puff at the entrance as well as at the top will stop that nonsense immediately.

When the beginner has smoked the bees and waited a while, he should remove the roof gently and lay it on the ground, with the bottom upwards, just behind the hive. If the colony has not yet got supers on, the next job is to remove the crown board. This is done by gently inserting the flat blade of the hive tool between the crown board and brood chamber at a corner of the hive and gently levering upwards. As this is done smoke should be puffed into the enlarging crack between crown board and brood chamber so that it drives the bees back, and prevents them coming out to see what is there. The crown board may lever up quite easily, particularly when the equipment is new, but as the bees propolise the joint, and where the propolis is old, the crown board will be well fastened on. In this case too hard a leverage from one corner will cause it to suddenly crack away, jarring the hive and the bees. This should be avoided by gently levering at each corner one after the other until the board is completely released.

Left *When lighting the smoker, make sure it is going well and producing cool smoke before you take it to the hive.*

Right *Let the cool smoke drift into the hive under the crownboard. Do not pump it in.*

The loose crown board can now be raised, puffing smoke under it as this is done. With the crown board now completely removed in one hand the beekeeper should drift smoke across the top bars of the frames until all the bees have gone down into the beeways between the combs. He should now take a quick look at the underside of the crown board to make sure the queen is not on it, and place it down below the hive entrance so that the bees still on it can make their way back indoors.

By the time he has done this the bees will be coming back up from the face of the combs on to the top bars again. The ones that just walk around on the top bars are usually quite inoffensive but it will be noticed that some stick half out from the beeway between the frames and that these, with their front legs in the air, will swivel to follow the movements of the beekeeper. These are the bees who may try to defend the colony, but a puff of smoke will send them below again out of the way. Repetition of this smoking down procedure should give the beekeeper control of the colony for the whole manipulation.

The first frame or the dummy board, if one is used, should now be released from its fellows with the crook of the hive tool and gently withdrawn. When using short-lugged frames they should be picked up by the top bar and as they are raised clear of the brood chamber wall the fingers crooked under the lugs. The use of a little smoke may be necessary just to clear the bees away from under the fingers. Frames should be raised slowly and carefully. Bees on the face of the comb should not be rubbed against either the side wall or bees on the face of the next comb; they get a bit annoyed at being rolled over one another. Nor should the side bar of the frame be allowed to touch the side wall of the brood chamber as there are likely to be bees there which would be crushed. Crushing bees, of course, releases the smell of venom and the pheromone which excites bees to sting. The old beekeeping saying that 'the first sting is the most expensive one' has quite a lot of meaning.

Once the first comb has been removed and examined it can be stood down in front of the hive near the crown board or, if you prefer, placed in a box carried for the purpose. It should certainly be left out of the hive until the examination is at an end because this will give more room for the removal of subsequent frames without 'rolling' the bees. Smoke should only be needed at times just to clear the bees away from the hive tool and the fingers or if they begin to get a little excited.

If the colony being examined has supers on it then the hive tool should be inserted between the bottom of the supers and the queen excluder, the supers levered up and smoke puffed into the opening. The supers should then be picked up and placed on the upturned roof behind the hive. Normally the bees will stay in the supers and not bother the beekeeper but sometimes, particularly in July and August, they may need another puff of smoke to keep them quiet. The queen excluder should now be given a gentle puff of smoke—do not use much because it sometimes annoys bees who are trying to struggle through the slots down out of the way. Loosen the excluder from the brood chamber and gently raise, puffing smoke under it, as you do so. Look to see if the queen is on it, and if she is not, place the excluder down in front of the hive by the crown board. If the queen *is* found on the excluder or crown board, run her back into the hive, or better still run her into a match box with a couple of workers and then when you have a frame containing brood out during the examination release the queen on to it. It is always safer to place a queen on to brood where the bees expect to see her than to run her in from the top bars or the entrance.

Once the examination is completed, the hive should be quietly and gently reassembled, using smoke to clear the bees away from places where they may be crushed. Care should be taken to align the boxes one above the other, as this will prevent escape of bees from places other than the entrance, and will not encourage excessive propolisation of the parts nor provide ledges which will cause rain to get in between the boxes.

Always work steadily and methodically. Avoid rapid, jerky movements and jar the hive as little as possible. Never crush a bee if it can be avoided. Use as little smoke as possible consistent with good control of the bees.

Colonies can be opened for examination at any time when the

temperature is over 17°C (62°F). If examination is essential below this temperature it should be performed as swiftly as possible, and only the bare essentials dealt with. Below 10°C (50°F) I do not advise the beekeeper to open up at all, and if he does he must be aware that he is putting the colony at risk. Brood may be chilled and the colony will need to eat more stores to bring their environment back to normal. It would be better left alone.

What to look for

Every time you open a colony you should ask the five questions. They are vital and should be memorized.

1 Has the colony sufficient room?
2 Is the queen present and laying the expected quantity of eggs?
3 **a** (early in season) Is the colony building up in size as fast as other colonies in the apiary?
  b (mid season) Are there any queen cells present in the colony?
4 Are there any signs of disease or abnormality?
5 Has the colony sufficient stores to last until the next inspection?

These have been mentioned already and will be examined in more detail in this chapter. In addition, you should keep an eye on certain practical matters. Hives should be sound and waterproof; any holes, either those which allow the bees passage at places other than the entrance, or those which allow water in, particularly in the roof, should be repaired. Stands should be examined to see that they are strong and stable. Brood combs should be watched for any increase of drone comb, and this replaced wherever it exceeds about 6 per cent of the area of the comb. Research has shown that colonies do not make more drone comb when they already have a fair amount in the hive. It is therefore good practice to leave at least one comb with more than 6 per cent—say up to 12 or 15 per cent—and place this at the edge of the brood to help reduce the amount made by the bees. The bees make the drone comb by tearing down a patch of worker cells, usually in the corners of the comb, and rebuilding in the larger 'drone' size. It will be found that, using this criterion, about two brood combs will need replacing each year. Combs which contain mouldy pollen will also need replacing (see page 210). Such pollen turns into a hard mass which cannot be broken up by the bees—the only way the bees can remove it is by tearing the cells down to the septum and removing the pollen in cell-sized lumps.

The presence of too much drone comb or of mouldy pollen are the only reasons why combs need replacing, providing the standard of beekeeping is adequate and the colonies good. Combs which have dried out and partly mouldered away, or which have large holes in them, should also be replaced, but these will not occur if large well-fed colonies are kept in sound hives.

Assessing the queen

In order to answer question 2, the beginnner must quickly learn to assess the queen, not as a representative of a particular strain of honeybees as compared with queens from other strains, but as a productive unit, either worth leaving in the hive or past her usefulness and due to be replaced.

First we must look at her egg laying and discover whether she is increasing the size of her broodnest, holding it static, or reducing its size. During the early part of the season when the colony is still building up she should be re-laying cells almost as soon as they become empty on the emergence of the occupants. She should also be expanding her broodnest area both on combs within the broodnest and extending on to adjacent ones. If a queen is laying the same number of eggs each day, the ratio of the brood will be the same as the length of the stages of the life cycle: 1 egg to 2 unsealed larvae to 4 sealed brood. In other words, one seventh of the area of the broodnest should contain eggs. Where a queen is increasing her egg-laying rate this ratio will be reduced, I would suggest to about $1\frac{1}{2}:2\frac{1}{4}:4$, and thus one fifth of the area should contain eggs.

The next thing to look at is the pattern of her brood. Sealed brood areas should be completely sealed over, with very few empty cells or cells containing young larvae. A lot of empty cells means that a lot of the larvae are dying off, which may be due to the age of the queen or her quality. I would not want to see more than about 5 per cent unsealed cells, that is about five in any 2 inch square of sealed brood. Poor or old queens may produce up to 50 per cent non-viable larvae. Examination of open cell areas may show great disparity of larval age in adjoining cells, indicating that larvae are dying at an earlier age. The queen may lay up every cell in an area of comb, but deaths occurring in the larval stage will spoil the pattern.

Providing the egg-laying rate and the open and sealed brood patterns are acceptable, and the queen is expanding her laying in spring then she is worth leaving to continue the good work. When the colony is fully built up and the queen is using most of the brood chamber comb area she will no longer be able to increase her egg-laying rate, but she should continue to re-lay cells as soon as they are empty. This means that she is not reducing the size of her broodnest and can be assessed upon her ability to keep going and on the pattern of brood. Not until mid July, certainly in the south of England, should she begin to allow her broodnest to start to retract, and even then the rate at which she reduces the size of it should be fairly slow.

A good comb of sealed brood with very few empty cells, showing highly viable brood.

Queens can fail and become uneconomical at any age and at any time of the year. The speed at which queens fail can vary very considerably. Some will fail very rapidly, going from first class egg production to laying not more than a few dozen eggs a day in the period between two inspections. On the other hand, many queens will begin to fail very slowly and the reduction will not be noticed for several examinations. This, if it happens early in the season, will reduce the value of the colony as a honey production unit, and will require special effort to bring it back to first class condition. The skilful beekeeper will recognize a failing queen early and replace her before too much damage is sustained.

Reduction in egg laying can also indicate that the colony is producing queen cells and has made up its mind to swarm. This possibility should always be checked and the appropriate action for dealing with swarming colonies set in motion (see Chapter 7).

Many queens reduce their laying because one of their back legs becomes stiff and paralysed. This always seems to hinder them very considerably and they never, in my experience, recover. A defective queen should therefore be replaced immediately, as should queens who have a paralysed front leg, which causes them to be superseded. Replacement saves time and the mishaps which can occur if supersedure is allowed to take its course in mid season. Queens which have a paralysed mid leg do not appear to be incommoded in the slightest. The leg dries out and becomes 'polished' in appearance. I always feel it must be a nuisance and a possible entry point for infection and therefore would snip it off close to the body. This operation does not appear to be noticed by the queen, nor does it, as far as I can see, affect her length of life.

Assessment of the colony

The queen is the mother of the colony and therefore all its characteristics come from her. It is possible to have a queen laying the right proportion of eggs and with a perfect brood pattern, but producing a colony less than half the size of the rest of the colonies in the apiary. If left to its own devices this colony would probably have little honey at the end of the year, while the others are producing a reasonable crop.

The two main reasons for small colonies in the spring and mid season are infection with nosema and a poor queen. A poor queen may be young and laying consistently, but incapable of laying in sufficient quantity to produce a productive colony. This lack of quality may be due to her being of a poor breed or a non-productive strain, but more probably she was not fed adequately as a larva or was converted from a worker larva to a queen larva too late in her developmental period. A little colony led by a failing queen will be observably different from a nosemic colony. In the 'poor queen' colony the worker bees are living their full length of life and it is the queen who is holding the colony back. The picture one gets in the broodnest is that there are plenty of bees, the brood is well covered with workers and many of these are standing about or working on the empty and store combs. In the nosemic colony, on the other hand, the workers are dying early, and if the queen is not affected with nosema herself, she is laying as large a broodnest as the number of workers can look after. The picture here, therefore, is one of a broodnest very sparsely covered with bees and few, if any, on the empty and store frames. Few queens themselves suffer from nosema and if they are infected they fail and die very quickly so that the problem is made more obvious.

If you correctly diagnose one of these causes for small colonies, you must act. Poor queens should be requeened and nosemic colonies treated as described in Chapter 9. Other reasons for small colonies in the spring are usually, by the time you see them, past history. Weakness may result from their having been put into winter in leaky hives or short of stores, sited in wet or very exposed positions, perhaps because hedges have been removed, or may be the result of an early shut down of queens the previous autumn owing to wet, miserable weather. Infection with some of the virus diseases could be a contributory cause. In all such cases you can only learn from your mistakes where you are at fault and resolve not to make the same mistakes again. Sometimes it is beyond the beekeeper to foresee the problems, and even if he could there would be nothing he could do to stop them. He must return to the building-up process and get the colonies back to size as rapidly as possible.

Building up small colonies

Once the cause of the smallness has been removed the colony wants one basic thing: more population, more worker bees. These can be taken from a colony which is doing well—the big colony that can lose a bit of brood or a few bees and hardly feel the loss. Care should of course be taken that you are not moving disease around but the big colony will usually be a healthy one.

Look at the small colony and see how many bees it has surplus to those looking after brood. If there are plenty, then a comb with a small patch of sealed brood could be given to it. If on the other hand it has few surplus bees then what it urgently needs is extra workers, adult bees which can look after themselves, not brood which still needs keeping warm within the cluster. How would bees be added to a colony? I use the following method. Go to the small colony and open up, remove one or two of the empty or store frames to give a space at the side in the brood chamber. Put the crown board back on and go to the large colony, smoke and open up. Go through rapidly and find the queen. Put the frame with the queen on it in a nucleus box, or put the queen in a match box with one or two workers. The match box could, incidentally, be put in the hive entrance so that, should you forget about her, the bees would release her. Go swiftly through the large colony and find a comb on which worker bees are emerging, hatching from the cells. Give this comb a moderate shake. You will learn with experience that as you shake with increasing power so progressively younger bees drop from the comb, so a gentle shake removes the oldest bees, a moderate shake removes the old bees and a lot of the house bees, a very heavy shake will remove all the bees except those just hatched. These will have to be picked off individually if you want to get rid of them, but in the current manipulation we want young bees so a moderate shake will do. The frame is then carried across to the small colony and the bees are all shaken from the comb onto the floor of the hive. These bees will submit when challenged and will become members of the colony within a couple of hours. The comb is then returned to its own colony and the queen released. In this way a small colony can rapidly be given extra population which will bring it to a strength where it has bees in excess of those needed in the broodnest. From then on the colony can be given small areas of brood and then larger ones as the population begins to reach a normal size for the time of the year. When these colonies reach normal size future build-up can be assisted by the method of spreading the brood as described in Chapter 5.

Supering

Colonies requiring extra room are usually given shallow 'supers' although some people work with boxes all of the brood chamber size. The disadvantage of this is that a brood chamber full of honey is a very considerable weight and more than one would wish to lift about. Beginners may be confused because beekeepers often call the deep boxes 'brood chambers', and the shallow ones 'supers', irrespective of the job they are actually doing at the time.

The general rule for supering is that the bees should never be using all the comb available to them. As soon as they get near this state a super should be put on, but remember that the aim is to draw bees from the brood chamber into the super fairly quickly. The beginner will only have foundation in his supers, and bees will often not go quickly through a queen excluder to get to a super of foundation. Thus I would put the super on *without* an excluder. At the next inspection the bees should be established in the super, and be drawing out the wax into comb. The queen can then be found and if she is in the super be put down into the brood chamber and the excluder put in place beneath the super.

Foundation in a super should be spaced at $1\frac{1}{2}$ inch centre to centre by using narrow metal ends or castellated runners (see page 73). As soon as the combs are drawn out to the usual $\frac{7}{8}$ inch thickness a couple of combs can be removed and the spacing increased to 2 inches, now using *large* metal ends or appropriate castellated spacing. This of course cuts down the number and cost of the frames in the super for the same amount of honey. The bees will continue drawing out the comb until there is a single bee space between the faces of two adjoining combs. These fat combs of honey are much easier to decap (see page 242) when extracting. Two stages of spacing are needed because if large spacers are used with foundation, the bees are likely to build their own comb inconveniently in between the foundation rather than to draw this out. The whole problem is avoided by using 'Manley' super frames (see page 72), which are self-spaced at $1\frac{3}{4}$ inches and can be used at this spacing both with foundation or drawn comb.

After the first year the beginner should have some drawn comb and should mix this in with foundation in the supers. The drawn combs should be placed on the outside against the box wall, while the foundation is kept in the middle of the box where the heat from the brood chamber is greatest. This arrangement encourages the bees to enter the super and the warmth gives those pulling foundation considerable help. In fact it is worthwhile taking advantage of this

Workers drawing wax foundation. The depth of cells decreases from right to left.

distribution of heat in the box when the bees are dealing with the first boxes full of foundation. They will usually start to pull the foundation in the warmest part of the box, over the top of the broodnest. As they pull the combs out to a full $\frac{7}{8}$ inch depth of cell these drawn combs can be moved to the outside and the foundation moved in to the centre.

Even when you have all fully-drawn supers it is probably worth putting all first supers on without a queen excluder to get the bees quickly established in them, the excluder being put in as soon as this happens. By this method surplus bees will move quickly from the broodnest into the supers, thus relieving any incipient crowding in the former. This should help to prevent some colonies and to delay others from embarking on queen cell production. The disadvantage is that some of the queens will nip up and lay eggs in the supers, which will not matter at the time if the combs are all worker size but can be disastrous if the supers contain drone comb, which is why I never recommend the use of drone foundation in supers. In addition, combs which have been used for breeding provide much better food for wax moths than plain beeswax. The careful beekeeper will thus mark the 'first' super as such and keep it for this purpose each year. I think there is an even more rapid movement of the bees into these supers because they have been bred in in the past.

Winter storage of supers is dealt with in Chapter 9, on page 208. Method of storage will affect the way the supers are put on the next year. If supers are put away wet from the extractor, and stored over winter in this state, they will be sticky with honey at the time they are

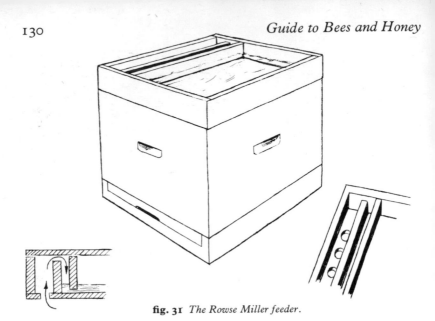

fig. 31 *The Rowse Miller feeder.*

put on in the spring. The reaction of bees who find honey which has
suddenly arrived in the hive is to dance and stimulate others to rush
out of the hive, causing quite a commotion and possible robbing. If,
therefore, you put wet supers on as you manipulate the colonies you
will be quickly surrounded by excited bees. It is better to mark the
colonies requiring supers and to put all of these on at the end when
routine examinations are finished. My method is to leave the roofs off
the colonies needing supers and then, when everything else has been
finished, to put a super by the side of each. The crownboard is then
removed from each hive and put straight on to the top of the super,
which is picked up and put on the hive, and adjusted carefully into
place. The roof can be put on when everything is finished.

Feeding

Bees should be fed with white granulated sugar mixed with water to
make a syrup. Brown sugars or raw sugars should not be used as these
are harmful to the bee, particularly as winter stores. The strength of
the syrup should be 2 lb. of sugar to every 1 pint of water. Sugar syrup
need not be boiled but may be made with hot water from the usual
household system, stirred until the crystals have dissolved. An easy
way to arrive at the correct strength of syrup without having to weigh
the sugar is as follows: take any container, half fill it with water and
then add sugar to fill. You will need 16 lb. of sugar to make 2 gallons of
syrup which will weigh approximately 26 lb.; when fed to the bees this
will produce about 23–24 lb. of stores equivalent to 20 lb. of honey.

When feeding at any time I would give the syrup to the bees as

The plastic bucket feeder is very good for contact feeding. The bucket can be half filled with water, and filled to the top with sugar. It can then be inverted without any mixing.

rapidly as possible so that they can take it down and store it where they want it, sealing it over as they would do with honey. Syrup should be fed to bees in one of the many types of feeder sold for the purpose. My own preference is for the Miller-type feeder, and particularly the design used by Mr David Rowse in Hampshire. This feeder is shown in fig. 31. Its advantage over the more usual type is that the place where the bees come up to feed is on one side; this means that should the hive be slightly out of level the feeder can be placed so that the feeding side is at the lowest level to avoid waste. When the bees have removed most of the syrup they can enter the main body of the feeder and clean it up completely, thus preventing the problem of taking off and packing away sticky feeders. Made about $3\frac{1}{2}$ inches deep, these feeders will hold about $2\frac{1}{2}$ gallons of syrup, and autumn feeding can therefore usually be accomplished with one feed.

Round aluminium feeders as shown on p. 93 are quite efficient, the only objection being that their small size necessitates several fillings for autumn feeding. This is probably no disadvantage for the beekeeper whose colonies are near at hand, but they are useless when dealing with out-apiaries.

Where Miller or aluminium feeders are used a small amount of syrup should be poured down the holes on to the bees after putting the feeder on to tell them there is sugar available above. Otherwise a colony may fail to find the syrup for several days, as sugar does not appear to have any smell which they recognize as food.

Plastic bucket feeders as shown above are useful and efficient for contact feeding in winter but have the disadvantage that a box is needed to surround them, for if the roof is balanced on the feeder without support it can easily be blown off. The feeder, filled with normal syrup made from 2 lb. of sugar to 1 pint of water, is put directly on the top bars of the frames, *upside down* (they have metal gauze in the lid), inside an extra brood chamber and with a sack packed around it to help keep the syrup warm.

Watertight tins with a dozen or so holes about $\frac{1}{16}$ inch in diameter punched in the lid are easy to adapt and often used. They are used in the same way as the plastic contact feeders, upside down on the frames. They have disadvantages, however, because if there is a large temperature change between night and day, with a very rapid warm-up in the morning, the air above the syrup in the tin expands so quickly that the syrup is expelled at the bottom faster than the bees can cope with it, and the result is sugar syrup running from the hive entrance—both a waste of sugar and likely to set up robbing by other colonies in the apiary.

Only pure syrup should be used, and in most years without additions. It should not be combined with a treatment or a preventative for disease, except for nosema.

Should you be making syrup which is to contain Fumidil 'B' in order to treat nosema (see Chapter 9), it will be found that the Fumidil powder is very fine and it is almost impossible to stir it into ready-made syrup. My method of mixing this substance is to take a large container and put 8 lb. of dry sugar into it. The bottle of Fumidil is now emptied into the container and the powder mixed into the dry sugar until it can no longer be seen. Four pints of warm water are now added and the whole stirred. It will be found that the Fumidil will be automatically dissolved with the sugar and will not float to the top. The usual small bottle of Fumidil powder contains 0.5 g. which is enough for three colonies, and should be fed to them in a total of 42 lb. of sugar. It will thus be necessary to reduce the concentration of Fumidil for the above mixing to the right strength by adding a further 34 lb. of sugar and 17 pints of water. You will end up with 42 pints of syrup which can be split up into three containers, each containing 14 pints of syrup and a third of a bottle of Fumidil. Each of these containers will constitute a dose for one colony. More will be said about this under the heading of disease.

I would not feed Fumidil every year, but try to monitor the incidence of nosema in the apiary and treat accordingly. I would certainly not use a treatment for any other disease, particularly AFB and EFB. In my view this is unnecessary except in very special cases and under special circumstances in which the effect of any treatment will be very carefully monitored by people competent to do so. Routine treatment for these diseases could, particularly in areas where pockets of high incidence of these diseases occur, cause considerable harm in the long run by masking the disease and by selecting out resistant strains of the causative bacteria.

Autumn feeding has been dealt with on page 99. Stimulative spring feeding of large colonies is rarely practised today as it has been shown to be a waste of time, having practically no effect. It is still used on small overwintering nuclei which are often in need of extra food by the beginning of March. These may also be helped by the water content of the syrup which reduces the amount they need to fetch in from outside.

Summer feeding should only be practised where the colony would starve without it: dead colonies get you no honey. Therefore if at any time during the year starvation of a colony is possible, or should the weather prevent any flight for ten days, then the colony must be fed. I would give them a gallon (8 lbs sugar) and hope that the weather would change before they had eaten it all up.

Moving colonies of bees

The old rule governing the movement of honeybee colonies is as valid today as it ever was: 'colonies may be moved under 3 feet or over 3 miles.' It is mandatory during the active season, when bees are flying most days. The reason for the rule is fairly easy to find: bees learn the district over which they fly and home on to their hive with complete accuracy, providing the picture of the surrounding area is not altered, as explained on page 37. The shift of over three miles is always necessary in the active season. If a move of, say, two miles is made, then as soon as the bees fly out half a mile they come across their old known flight pictures and fly home to their former site. A distance greater than three miles may be preferable where the colonies are being moved up or down a narrow, high-sided valley, as their normal flight patterns may extend over greater distances in situations of this sort.

Winter moves can be very much smaller, and after a week of frost when no flying has occurred colonies may be shifted about in the same apiary without much fear of their getting lost when flying begins. Colonies which are to be transported must be shut in when all the bees have stopped flying for the day. However, they will overheat and die if the entrance is closed without giving added ventilation, and therefore colonies for transporting are given a ventilated screen over the whole top of the hive, from which heat can readily escape. Large colonies shut in without a screen will very rapidly build up sufficient heat to reduce the tensile strength of the beeswax to a point where the combs will collapse, and honey will be released over everything. The bees themselves are turned dark in colour, and those that remain alive are unable to fly, so the colony soon dies. It is a very sorry sight and always happens to the big, prosperous and vigorous colonies first.

As the normal beehive is made up of separate boxes standing one above the other, these must be fastened together for transport in a way that is secure. Various patent clips are available but these need careful fixing to the boxes with a jig so that everything is interchangeable. Various strappings have been used, using steel or nylon bands, but I have never really trusted them as a box has only to twist about $\frac{1}{2}$ inch to let bees out. Bees escaping from hives on the move is part of beekeeping and provides many a tale and many a laugh afterwards, but I can manage without the excitement at the time. I therefore like hives well stapled up with little chance of falling apart, and I still prefer the double pointed nail or large box staple as shown in fig. 32.

The procedure for getting colonies ready for transport is as follows. During the day the colonies should be examined in the usual way, making sure the frames are well packed together so that there is no possibility of movement. The ventilated screen is put on and screwed down, and the entrance prepared for shutting in. I normally use one of two methods, both of which are easy and efficient, to close hives and shut the bees in. One is to use entrance blocks which have an entrance on one side only (see page 68). These can then be turned over and pushed into place to shut up the door. The other method is to use 18 inch long by $1\frac{1}{4}$ inch square plastic foam strips. These can be pushed into the entrance and do not work their way out when travelling. The entrance block is removed, and the foam strip pushed in about three quarters of the way across the hive—the end will kink and stand out at right angles and can easily be pushed in later, at the time of loading when the bees have finished flying. The roof should be put back on, on top of the screen, until the time comes to load up.

When flying for the day has ceased the entrances are closed and the roofs removed to allow heat to escape. The hive is then nailed up with staples or banded. When using staples they must be put in at an angle, as shown in fig. 32—the boxes can twist on each other if the staples are put in at right angles across the joint between boxes.

Once the colonies are prepared in this way they are loaded up on to a truck or whatever vehicle is used, with as little fuss as possible, and always with combs running fore-and-aft of the direction of travel, to prevent comb-slap.

Where long journeys are necessary and the bees have to remain shut in for considerable periods, they should be examined every hour or so. If the colonies are producing a loud roaring buzz a couple of cupfuls of water should be poured through the screens; the bees will suck this up and become much quieter and more contented. A little consideration of the colonies in this way will reduce the damage done by subjecting bees to the unnatural and stressful conditions of moving.

On arrival at the out-apiary, the hives should be set up and the bees released as quickly as possible. They will sometimes pour out with vengeance in their minds, while at others not a bee will move and no bad temper will be shown. It is as well, therefore, for the beekeeper to

fig. 32 *A hive prepared for transport, complete with screen and nailed up with crate staples, which must be put in at an angle.*

prepare for the former reaction and get everything ready for a rapid withdrawal. Roofs and crown boards should be laid out, one to each hive. The beekeeper then pulls the entrance block out and immediately puts the crown board in place on top of the screen and the roof on, and proceeds to the next hive. It is always done in this way because it is said that colonies have been known to die if the crown board is put on first and the hive opened afterwards. The crown board needs to be put on quickly, however, for if the bees rush out a lot of them may be attracted on to the screen board where they can smell their colony. They are then very difficult to get rid of and considerable time can be wasted trying to put the crown board in place and the roof on. The screen is left to be removed when the next examination is due, by which time the bees will have forgotten all about their journey.

Hive closure at the entrance should always be total. If it is done with perforated zinc or something else that the bees can see through many will kill themselves trying to get out. I used perforated zinc for some years and when it was in position antennae protruded from every hole as bees struggled to get out, and there was always a handful of dead bees on the floor afterwards. This does not occur if light is totally excluded.

If colonies are being moved from an apiary and returned within a fortnight it is best to number the colonies and sites so that each can be returned to its former place. Bees remember their old sites over this length of time, and quite a bit of drifting and fighting occurs on their old sites if the colonies are mixed up. After ten to fourteen days this ceases to happen, probably because most of the original foragers will have died during the stay at the out-apiary.

7 Controlling swarms and making increase

A colony that has produced queen cells or even fully developed queens does not necessarily have to swarm. Many will kill these queens or queen cells, giving up the whole process of swarming. In some colonies, of course, the new queens will supersede but this usually happens either in the beginning or, more generally, at the end of the active season. In the middle of the year colonies usually either swarm or give up the whole idea. No one as yet has been able to discover a method of differentiating between the colonies which will swarm and those which won't. The practical beekeeper therefore equates summer queen cell production with swarming and deals with the colonies from this angle. I shall continue to use the normal beekeeping parlance and write of a colony making queen cells as a 'swarming' colony, although the whole idea of this chapter is to help you prevent the colony actually coming out of the hive as a swarm.

Swarm prevention or delay

As the production of queen cells is mainly, if not entirely, controlled by the age of the queen and the congestion of the colony, attention to these two factors will do much to prevent or postpone the start of swarming. The age of queens should be kept to a minimum, consistent with value of the queens and their economic length of life: I would suggest they should not exceed two full seasons in large production colonies. They should also come from a strain which is not prone to swarming. This will be difficult for the beginner, as normally obtainable queens carry no information on their characteristics at all: it is a long-term objective to keep in mind when breeding your own—see Chapter 8. Congestion can be prevented by correct use of supers and ensuring the bees take up as rapidly as possible the extra room given them by encouraging them into the super (see page 128).

Shading colonies from direct mid-day sunshine is said to hold

The well-protected beekeeper lowers a swarm into a straw skep.

swarming back, but it is also said to reduce the rate of spring build-up. If both are true I would prefer the early spring build-up and the slightly earlier swarming in most areas. If you are situated in an area where nectar flows are late then the factor of shading should be taken into account.

Dealing with the swarming colony

At some time during the season, as the beekeeper conducts his routine inspections he will find queen cells, and must then deal with the colony or probably lose a swarm and often the honey crop. When the colony is open in front of you is not the time to make up your mind about what you are going to do. That way leads to panic measures. For their first few seasons beginners should adopt a complete method put forward by an experienced beekeeper and stick to it. Do not try, in the first few years, to combine bits from various people's methods, as often they are not compatible. Once you have been keeping bees for a few seasons and have begun to get an understanding of and a feeling for them, then experiment by all means. Who knows—you may make the great break-through in the handling of swarming.

In the meanwhile, may I suggest two methods which I find simple, reliable and least destructive to the honey crop. These are the *artificial swarm method* and the *requeening method*. The first method can be used by any beekeeper, the second only by one who is producing queens for his use early in the year.

Artificial swarm method

To carry out this method the beekeeper will need to have an extra brood chamber, floor, crown board and roof. The brood chamber should contain its ten frames with full sheets of foundation or, preferably, drawn comb.

Routine examinations of the colonies are carried out at weekly intervals, this being convenient to most beekeepers. Once colonies have built up and no further work is needed other than the provision of space and swarm prevention, then the amount of routine disturbance to the colony can be cut down. If a colony is not making queen cells then it can safely be left for fourteen days, and one weekly inspection can be missed. If the colony starts queen cells immediately the beekeeper leaves, it will not have a queen emerging from a cell for sixteen days, as first batches of queen cells for swarming are very, very rarely started on young existing larvae.

When queen cells are found during the routine examination action should be taken immediately to produce the artificial swarm. The supers will have been removed at the start of the examination. The brood chamber, on its floor, should now be lifted and placed about 2 feet away from its original site. A new brood chamber and floor are put

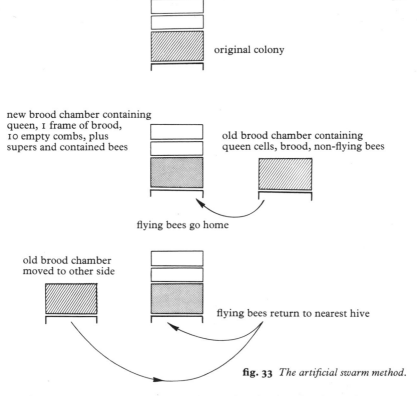

original colony

new brood chamber containing
queen, 1 frame of brood,
10 empty combs, plus
supers and contained bees

old brood chamber containing
queen cells, brood, non-flying bees

flying bees go home

old brood chamber
moved to other side

flying bees return to nearest hive

fig. 33 *The artificial swarm method.*

on the original site. The old brood chamber is examined and the queen found. She is then put, on the comb upon which she was found, in the centre of the new brood chamber on the original site. Any queen cells on the comb with the queen should be destroyed. The new brood chamber is then filled with ten, preferably drawn, combs but foundation will do if drawn combs are not available. The queen excluder is put in place, the supers replaced and the roof put on. This hive now contains the supers and the bees in them. The flying bees will of course return to their old site and join the queen. The population and organization of this hive is such that it is very like a swarm and should get on with the job of making a full colony and give up making queen cells.

The old brood chamber which is now a couple of feet from its original hive, with its entrance facing the same way, is examined and all sealed queen cells are removed, providing there are some unsealed queen cells in which the larvae are almost fully fed and ready for capping. A crown board and roof are put on the hive and it is left for a week. At the end of a week this brood chamber is moved to *the other side* of the original site on which now stands the artificial swarm. The

fig. 34 *Uniting the two brood chambers by the paper method after the new queen is mated and laying and the old one killed.*

result will be that all the workers that have learnt to fly during the week, and there will be quite a number of them, will return to their last site and from there to the *original* site, thus further augmenting the population of the artificial swarm.

It is in order to be able to do this move of the old brood chamber without the fear of a young queen flying from it that sealed cells are killed when the colony is first split up. Queen cells are sealed for eight days, and therefore with no sealed cells there can be no virgin queen to lose her bearings when the switch is made at the end of seven days. The old brood chamber can be left alone after this until the new young queen has emerged, mated, and started to lay. Usually there is no need to go through it to remove all but one of the queen cells because the drastic reduction of population will cause the bees to give up any idea of swarming and will destroy all but one themselves.

It is important to ensure both colonies have sufficient food. This is particularly likely to be a problem with the old brood chamber, the combs of which may contain very little in the way of stores, as these were kept in the supers which are now on the new brood chamber. Feeding the colony is the answer, and I would give them a gallon of syrup in a rapid feeder.

At the next manipulation the colony on the old site should be examined to see that the old queen is laying up the empty combs and that no queen cells are being made.

Once the new queen in the old brood chamber has mated and started to lay, her colony can be united with the original colony after the original queen has been found and removed. To unite the colonies, a sheet of newspaper is placed on top of the supers and held down by means of a queen excluder. A few holes should be pricked through the paper with a pin or the corner of the hive tool. The old brood chamber containing the new queen is then put on top and the whole hive closed down and left alone for a week. The bees will chew their way through the paper in a few hours, and the time delay accustoms them to one another without fighting. At the end of this period the top brood chamber is placed down on the hive floor, after the bottom brood chamber has been moved to one side. The brood chamber on the floor is now made up with brood from the other one until it contains eleven frames of brood, or all the brood from the two boxes is used up and made up to the eleven with empty comb. If there are more than eleven frames of brood in the two boxes then the extra brood can be given to other colonies, or put back on top of the newly assembled colony where it is left until it hatches out. In this case I would put the oldest sealed brood in the top box and fill the space around it in the box with a couple of sacks to prevent the bees building comb in it. I do not like putting the brood chamber back on top full of comb, to be used as a super, because it is so difficult to uncap old comb.

This method accomplishes the two essentials of any swarm control system: it stops the colony swarming out, and replaces the queen. The latter is necessary, for if she has tried to swarm this year she will certainly do it again the next year.

The method as described does not make any increase in the number of colonies kept by the beekeeper. If he should want to make increase as well, then the method can be modified to provide it, but at the loss of some honey. The artificial swarm would be made in the same way as above, but at the end of the seven days when the old brood chamber is switched to the other side of the old site the following alterations could be made in procedure. The old brood chamber could be opened and a small frame of brood with a good queen cell on it placed in a nucleus hive. To this should be added a frame of stores and sufficient bees to look after the brood. This nucleus should be placed at the side of the artificial swarm hive opposite to that from which it was taken. No further interference would be needed until the new queen had mated and started laying, when the nucleus would be united with the artificial swarm after the old queen's removal. In effect this would become a queen-introduction nucleus and greater detail of this method is given on page 158. The rest of the bees, brood and queen cells in the old brood chamber can be put on a new permanent site in the apiary and, once the queen is mated, built up into a full colony by the usual means (see pages 105–11).

The requeening method of swarm control

The natural cycle of producing queens is described in Chapter 2, and the maximum safe period between inspections, assuming the resident queen has clipped wings, is ten days, as described on page 112. For requeening, it is assumed that the beekeeper has some form of queen rearing and has young mated queens available for use all the time.

Colonies are examined and the usual five questions are asked. Incipient queen cups are examined for eggs or larvae. As mentioned earlier, I would ignore a few eggs in queen cups, only increasing my vigilance in examining them the next time. It is very noticeable that eggs will be found in queen cups for several weeks before larvae are found in any of them, and I am always very doubtful as to whether these are the same eggs all the time.

As soon as a larva is seen in a single cell, shake the bees from the combs and search for and destroy all queen cells. This technique needs some explanation. Why *shake* the bees off the comb? The answer is because however experienced a beekeeper you are you will miss cells if you look for them with the bees still on the combs. Half a dozen workers sitting on a cell will completely hide it from view. There is of course no need to shake every bee off, but you must be able to see right across the comb. The combs are shaken into the hive so that the bees fall on the hive floor. My own method is to tuck a couple of fingers under the lugs of the frame and without removing the frame from the brood chamber rap the fingers on the edge of the hive a couple of times with a wristy movement. This dislodges the bees with very little movement of the comb, thus helping to cut down any chance of crushing bees between the side bar of the frame and the wall of the hive. Having removed most of the bees in this way the comb is carefully searched for queen cells, and all of these, including those with eggs in them, are destroyed. Care is needed to ensure that eggs, larvae and pupa in queen cells are killed, as bees will repair damaged queen cells containing a larva which is still alive. All the combs are carefully gone over in this way, after which the hive closed down until the next inspection, a note of the presence of queen cells being made on the record.

When the next examination of the colonies is made, some will have given up making queen cells and therefore only require the routine work of checking queen, stores, room and disease. Others will have made queen cells again and in these colonies careful note is taken of the amount of egg laying the queen is doing. If she is laying well, with hardly any reduction in her rate of re-laying empty cells in the brood area, then the colony is 'shaken through' again and all the new queen cells destroyed. On the other hand, if she is cutting down her rate of laying eggs, indicated by a considerable number of completely empty cells, then the queen should be found and removed, and all the queen cells destroyed. A nucleus should be made up, a new young laying queen introduced into it, and the nucleus placed beside the hive ready for putting into the colony next visit.

This process is repeated with all the colonies in which queen cells are found until either they have given up making queen cells or they have been requeened. It is uneconomic to shake through and destroy queen cells more than three times, especially where large full-sized or sealed queen cells are present on the second and third inspection—I would only allow them two chances before requeening. About a quarter of the colonies making queen cells give up doing so but unfortunately no one can find a way of telling which ones they will be.

In some cases the bees will have tried to swarm and have returned minus the clipped queen. If this has happened more than three days

A comb of brood from which most of the bees have been shaken while looking for queen cells.

before it will be obvious by the complete lack of eggs in the colony. Requeening can be set in motion by the nucleus method and all the queen cells destroyed.

If the queen has been lost within the last three days it is difficult to decide if she is gone or not, and much will depend upon the beekeeper's skill in interpreting what he sees in the colony. For the inexperienced it is probably best left for the next inspection to make the matter clear, but all the queen cells must be destroyed as usual before the colony is left. In many cases, of course, the beekeeper will not realize the queen has gone at all until the next inspection, when he will find no eggs and no young brood, and in fact he can calculate exactly when the queen was lost by the age of the youngest brood. A more experienced beekeeper may feel that the queen is probably gone, and without any definite proof he may then risk making up the requeening nucleus and introducing the new young queen to it, but he must eliminate as much as possible the risk of getting the old queen into the nucleus as he is making it up by careful examination of all the bees put in. If the old queen does slip through to the nucleus the new queen will certainly be killed by the bees.

Sometimes examinations have to be put off because of heavy rain, with the result that when the colonies are examined the old clipped queen will have gone and young virgin queens will be emerging from their cells. Some may have already done so. Often in a colony in this state the worker bees will be physically holding the young queens in their cells by clustering on the opening, in preparation for swarming.

When the beekeeper starts work they all leave the queen cells and in a few minutes the young queens will dash out of their cells. To save time, therefore, the beekeeper, as soon as he realizes the condition of the colony, can rush through destroying all the large cells. He should put some of the young virgins away in matchboxes, one in each, in case he finds he needs them later.

The experienced beekeeper will have been counting the hatches (see page 146 for technique) and will then find the young virgins and remove them. When he has found them all he can requeen with his own mated laying queen as described on page 159. In brief, the rapid destruction by the beekeeper of imminently hatching virgins will reduce the amount of work necessary to clear the colony of the unwanted virgin queens which would kill the beekeeper's introduced queen.

The less experienced will not be able to find virgin queens very easily, and therefore would be best advised to release a couple of young queens from their cells—'pull' them in beekeeping jargon—and then destroy all the other queen cells in the colony. No matter how many young virgin queens are left loose in a colony it will not swarm unless there is one or more queen cells left in the hive as well. This rule is a useful one as it can be used when in doubt as to what exactly is happening in the colony. The idea of leaving a couple of young virgin queens in the colony is that you will be quite sure that there are *some* queens left in the hive. The sight of one hatched cell is, in my experience, not conclusive and if no young queens are left this often results in a queenless colony.

Beginners will of course make mistakes in handling; colonies will be in the state of having queen cells and the beekeeper will not be able to decide what is happening, or why. Providing there are no eggs in the colony, thus indicating that the queen is gone, any colony can be repaired by leaving a good queen cell. The disadvantage of doing this as a routine method of dealing with swarming colonies is that the queens usually take about three weeks to mate and start laying. More importantly, during this period the colony will not work and, even when other colonies are storing honey in moderate quantity, will make almost no increase in weight, collecting just enough for maintenance.

The swarmed colony

Although I hope you will use one of the above methods to avoid swarming, you should know how to deal with a colony with an unclipped queen when a swarm does happen. Two different situations arise: the swarm is captured or it flies away and is lost.

When a swarm has been captured in the apiary it is necessary to be absolutely sure that it is your own if you are going to follow the method detailed below. Someone must have seen it come out of the hive if you are going to be sure. Alternatively, if you have all your queens marked (see page 157) with different coloured paint you can spot your own and know which hive she comes from. If you are able to find the queen in the swarm and take her away, the bees will start to move back to their home within twenty minutes, and you will know the source.

If the beekeeper knows for certain it is his swarm, it can be handled in the same way as the artificial swarm. The colony from which the swarm has come out is lifted about 2 feet to one side, a new brood chamber and floor is placed on the old site, the brood chamber being filled with the full number of frames containing full sheets of foundation. The swarm is put into the new brood chamber by one of the two methods detailed on page 151. The supers removed from the old brood chamber are placed, above a queen excluder, on the new brood chamber, and the hive is fully assembled and left for the flying bees to return to their old site. The old brood chamber and its contents can then be handled as it is during artificial swarming. In fact the result is much the same, but this is a real swarm with the normal eagerness to work and to build comb which the artificial swarm lacks. For this reason they can be given only foundation in the brood frames as they will draw it out into comb quickly and perfectly at little cost to the beekeeper. In the end the whole lot will be united again, the old queen destroyed and replaced by the new one in the old brood chamber (see page 140).

In the second case, where the swarm has been lost, the beekeeper must deal with the colony as soon as possible to prevent other swarms, or casts, from coming out as well. The colony is opened and a good queen cell is found and is left to produce a queen (some beekeepers mark the comb by putting a drawing pin in the top bar above the cell). On no account must the chosen cell be on a comb that is shaken, or damage may result to the queen, who is quite loose in the cell. The comb should be searched thoroughly to ensure that no further queen cells are left on it, and the other combs should be shaken through and any other queen cells destroyed.

If no hatched cells are found amongst those destroyed the colony is then left for ten to twenty days before being examined again, when the new queen will have emerged and should have mated and started to lay. It is often three weeks or more before a young queen will come into lay in a large colony. Do not be impatient and think the colony is queenless: it is unlikely to be so. The new queen is just slow in getting started. A more detailed understanding of this situation will be gained by reading the section on queenlessness in Chapter 9.

If in searching for other queen cells you find hatched and emerged queen cells, then there will probably be others. The beekeeper can act as midwife to one or two and 'pull' them, leaving these in the hive as more mature than the selected cell, which can now be destroyed with all the rest.

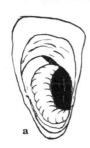  

fig. 35 *Queen cells.*

Destroying queen cells

When destroying sealed queen cells always make sure that none of them has hatched so that there is already a virgin loose in the hive. It is worth considering what happens to queen cells when they have hatched to make the technique clear.

When the queen cell is sealed, the larva goes on eating for about a day and then moves down into the pointed end of the cell and spins its cocoon around the last third, as above, fig. 35a. When the new young queen has finally moulted from pupa to adult she cuts around the pointed end of the cell until this falls down as a hinged cap, shown in fig. **b**. After the queen has gone, the cap can become totally detached and lost as in **c**. In this state it can often be confused with a queen cell in which the larva has died or which has become empty for some other reason. The difference is easy to test because in this latter case the cocoon will be missing, so the end will be very soft and a corner of the hive tool will pass through it easily. The hatched queen cell, on the other hand, because of the cocoon, is very tough and the corner of a hive tool pressed into it will deform it but not pass through easily.

The hinged cap on a hatched queen cell is often replaced by the bees and sealed on with wax, but because the cocoon has been cut this cap will come off at the slightest touch. Often when the cap is sealed on again it is done while a worker bee is inside eating the residue of the royal jelly. All of the hundreds of workers I have found in this position have been dead: the cell is too narrow for them to turn around so they are always head upwards towards the royal jelly.

When sealed cells are being destroyed, therefore, I first take the end gently with my thumb and first two fingers and tear it off the comb. Usually it breaks just over half way along its length. I then look at what I have in my fingers. Hatched cells will be empty or show the head of a dead worker as described. A wriggling tail will be a queen ready to emerge, and if you have done the job gently she can be pulled and used if required in the colony or taken away for use elsewhere. Finally, a white or light-coloured still tail will be a queen not yet moulted and therefore not ready for use by the beekeeper.

Left *A nice shaped queen cell showing the rough cellular surface.* **fig. 36** *a queen cell torn down by the workers.*

If cells are being cut out to be taken elsewhere for introduction to other colonies, keep them warm and put them safely somewhere so that if they do hatch while you are still working the queen cannot get back into the colony again—countless times when I first started beekeeping I put queen cells on the roof of the next hive only to find them hatched and gone by the time I was ready to pick them up.

Selection of queen cells

When selecting a queen cell to take over the colony, it should be chosen as follows. It should be about $1\frac{1}{4}$ inches long, shaped as shown above, broad rather than long. At least two thirds of the cell, on the side nearest the comb, should be well roughened with coarse ridges. Never choose a smooth cell as there is usually something wrong with it. I would prefer not to choose a cell which is totally surrounded by drone cells, as on occasions these queen cells can contain drone larvae. Finally, your chosen cell should be lightly touched on the point with the hive tool or fingernail to ensure that it is not an emerged cell with its lid fastened back on. If you wish, you can gently open up a flap on its side towards the base with a pocket knife, take a look at the queen pupa and then push the flap back and carefully repair the cut with the flat of the knife: you have to do a good job or the bees will tear it down.

Torn-down queen cells

If a colony has been left to swarm out several times they will reach a time when they will swarm no more, and any queen cells left in the colony will be torn down. The same picture will be seen in a colony which has been making queen cells and has decided to give up of their own volition. Queen cells which are taken from one colony and put into another colony will sometimes be torn down. In all these cases the torn-down cell will look like fig. 36. Where a colony is giving up the idea of swarming, unsealed cells may also change in appearance: the larvae will be removed—probably eaten—and the surface of the royal jelly will be covered in tiny pits, where bees have each taken a mouthful.

Taking a swarm

At some time or other every beekeeper will have to take a swarm, either his own or someone else's. Most beekeepers look upon taking swarms as a service to the general community which is usually rather, or very, afraid of them and glad to see them dealt with.

Swarms should be approached with your veil on and gloves if you wish. Do not take any notice of the old beekeepers who say swarms do not sting. *Never* try to take them without a veil. Usually they will be very quiet and co-operative, and you will have no trouble in collecting them. This is particularly so if they have just come out from a colony which has plenty of stores so that they are all full of honey, or if the weather is very fine, so that although they have been out several days they have been able to keep themselves topped up with nectar. But if they have come from a starving colony carrying very little stores, or have been hanging up for several days in bad weather and have used up a lot of their honey, then they can be quite nasty when shaken; fortunately one does not come across many swarms in this state.

Swarms are found in three types of position, each needing different treatment, but the technique of taking swarms is based on their behaviour pattern, which is to move upwards into the dark, and to stay there if the queen is with them. They can therefore be fairly easily persuaded to enter a skep or box. I prefer to use the old fashioned straw skep as illustrated opposite as the bees are able to hold on to it easily. It has some insulating properties which help them to keep cool once inside, and it is somewhat flexible and can thus be pushed into awkward places. A box can be just as efficient if it is firm enough to stand the weight of bees hanging from its top. Cardboard boxes are not too useful unless sturdily made and well stapled together, and they will become soft in the rain: I have seen more than one collapse under the weight of the bees.

The ideal position for a swarm from the beekeeper's point of view is on a thin whippy branch of a tree about 3 feet above the ground. The skep or box can then be placed under it, the branch firmly shaken, and the bees will drop off into the skep. I like to spread a large white sheet below the swarm before I shake them so that once in the skep the whole thing can be turned over on to the sheet, and the skep propped up on one side on a stone so that the bees can go in or out. If you smoke the remaining bees on the branch very heavily they will fly and most of them will be attracted to the bees in the skep, some of which will be fanning and scenting to call in stragglers. Using smoke in this way will often cause the queen to join the bees in the skep if you missed her when shaking. Providing the queen is in the skep the swarm will usually remain inside and start setting up house. If you have missed the queen they will begin to look for her within a few minutes, and once found they will join her again, probably back on the original branch.

If you are able to place a skep over the top, the swarm can be encouraged, with a little smoke, to walk upwards into the dark.

Having got the swarm in the skep, and waited twenty minutes or so to make sure you have got the queen and they are going to stay in it, it should be shaded from the sun and left for the bees to cease flying for the day, when it can be taken away to its new home. This is where the sheet comes in useful as you can tie the corners of this over the top of the skep, tie string around it so that bees cannot creep up the sides and escape, pick it up by the knots, and away home. I would not put it in the boot of the car as this may be hot and smelly with petrol; better to put it on the front seat beside you. Do not worry if one or two bees appear in the car; they will be too busy trying to get out to worry about you and once the engine is started the vibration will cause most of them to sit tight.

Most swarms are not in such an ideal place. Instead of being on a nice small branch they are often on a thick one, on the side of a concrete post, or even on the side of the house. In any case they cannot be shaken. They can, however, be invited into the box or skep by putting this over the top of them, as illustrated above. A puff of smoke will start them walking upwards into the dark, and if they are reluctant to go scoop a handful off the swarm and throw them up into the skep. Some will cling on and start fanning, and as soon as the scent reaches the others they will turn and walk into the dark like well-drilled soldiers. If you can put the skep over the top of a swarm, it can be taken in this way.

Finally, there is the swarm which is underneath something solid which cannot be shaken, nor a skep placed above it. I remember one swarm which was up inside the front wing of a car in the middle of a town. The only answer then was to put the sheet under the bees as much as possible, brush them down on to it, and then put the skep down touching them, propped up so that they could get inside. For brushing, a feather is best (or a goose wing if you can get one) as the bees get tangled in the bristles of an ordinary brush. By smoking one can coax them to start running into the skep, and once they are on the way time is usually all that is needed. Keep brushing them down if they try to climb up again without going into the skep: until the queen has entered the skep you have not succeeded in taking the swarm. She may have joined a cluster outside the skep unless you have kept them all on the move.

Swarms are not always on or near the ground and many are taken at considerable height in trees and on ledges. If you are prepared to have a go, remember it is dangerous and take precautions to cut the danger down. Be sure the ladder is standing firm, and not being held by a non-beekeeper. I well remember pouring half a swarm over a friend standing holding the ladder until I had reached safety—I must say he seemed quite excited about the ones inside his clothes. Others have not been so lucky and have had nasty falls. The thing that most surprised me when I first reached out to take a swarm was its weight when it landed in the skep. I did not appreciate—and countless other beekeepers tell the same story—that bees had weight, and the arrival of 6 or 7 lb. into the skep came as a considerable shock. So be warned: a good swarm can weigh up to 10 lb., quite a lot to arrive in a solid mass. Bees average about 3,500 to the pound, so you can work out approximately how many bees you have if you can weigh the swarm.

Hiving the swarm

Having taken the swarm and got them home you have to put them into their new hive. I would always give them foundation because they make such a good job of drawing it out. They can be hived in two ways, one traditional, the other quick.

The traditional way is most satisfying to the beekeeper, especially the beginner, but even the old timer cannot refrain from watching with delight. The hive is set up with the brood chamber filled with its full number of frames, each with a full sheet of worker foundation in it, and crown board and roof on. If a glass crown board or 'quilt' is used it must be covered, as the bees will only move into the hive if it is dark inside. A board about 18 inches wide and 3 feet long is placed sloping

The skep can be placed on a cloth or board, wedged up on one side, and left for the stragglers to find.

down from the entrance at an angle. The skep in its sheet is placed on this, the sheet untied and laid out flat on the board. The skep is then picked up and a hard downward shake throws most of the bees out on to the sheet. The rest of the bees can then be knocked out by banging the bottom edge of the skep against the hand. The bees will land in a large heap on the sheet and will begin to spread out in all directions, but mainly moving up-hill. As soon as the first bees find the hive entrance, and no doubt smell the comb, they will start to fan and scent. As the scent reaches the other bees they will all move to face the hive and begin to make their way up and in. Should the bees be slow to find the hive entrance a few scooped up and thrown into the entrance will start things off. The beekeeper can now sit down and watch what goes on, and look for the queen to see what she is like. It is always useful to know what colour the queen is because should you want to find her later she is easier to find if you know exactly what you are looking for. The bees will often take hours before they have all gone inside. They can be hurried, by smoke, but we have another method for those in a hurry.

The hive is set up for the quick method with the crown board off and completely empty of frames, but these should be handy, close by. It is a good idea completely to close the entrance. The skep is now untied, picked up, and holding it over the empty brood chamber the bees are poured and shaken into it. The bees will end up on the floor of the hive in a large heap. The frames containing foundation are now placed in the hive resting on the heap of bees. Do not force the frames down or you will kill some of your bees. The swarm will by this time be crawling up the foundation and the frames will slowly sink into place. Make sure you have the full complement of frames in place and adjusted correctly. Place the crown board on and open the hive entrance. The reason for shutting the entrance is that occasionally when a swarm is put in in this way the queen will fall near the entrance, which is also an exit, walk through it and take to the wing, with all the bees following. Shutting the entrance prevents this happening and she is unlikely to come out once they have moved on to the foundation.

In all cases when a large prime swarm has been put in by either method I would put a Miller feeder on the hive and give them a gallon of syrup to get them drawing the foundation, and to set them up in a prosperous condition. If the weather is poor for the next ten days I would give them a further gallon. If the weather is good they will probably be able to manage and may even start storing honey after the first two or three days. Large swarms need supers fairly quickly.

Many swarms you may come across will be after-swarms or casts. These will be headed by a young unmated queen, or in some cases by several. The first cast is usually fairly large, about the size of a football, but later casts can be very small, no bigger than a man's fist. Often there may be more than one of these small swarms together, or you may see them joining up and separating. In these cases there will be more than one queen in the little clusters and they will hop about all over the place. If put in a hive they will be out again next day or even the same day if they are put in fairly early. I can remember putting one into a hive five times in five days. I tried stabilizing them with drawn comb, stores and brood but they refused to stay, and on the sixth morning came out and flew away into the distance as fast as they could go. I waved goodbye and did not bother to follow. Many years ago, when trying to build up a number of colonies, I tried taking large casts with more than one queen and splitting them into two or three with a queen in each. I never succeeded in ending with more than one piece of the swarm with a queen which mated and started to lay. The other pieces would become queenless, even if the three pieces were taken to three different apiaries. Thus I have always felt that the bees had already decided on who was to die and who to live when they left the hive. Swarms of this sort are a liability rather than being of value but I usually collect them to keep them from annoying others. They usually get very cavalier treatment: I look for a really big colony with three or more supers on, I put another queen excluder on the top of the top super and an empty super on this. The cast is then poured into the empty super, the crown board put on, and the feed hole closed if there is one, and they are left to sort themselves out. Super bees are not very aggressive and usually they unite quite amicably but the queen or queens of such a swarm are found dead on the queen excluder at the next examination.

Finding the queen

So many of the manipulations for working colonies start with the instruction, 'find the queen', and as so many beekeepers find this almost impossible the whole thing breaks down immediately. The main problem of finding the queen is caused because so many beekeepers have such poor little queens in their hives. Properly produced queens are large and easy to see, particularly if the bees themselves are quiet on the comb, and not rushing about in all directions. The truth of this is brought home to me every season when beekeepers on practical courses easily see the queens in the demonstration colonies and remark, 'Why is it I can always see the queens here and not at home?' Therefore I would stress once more the necessity of producing good queens as described in Chapter 8.

With good large queens in your colonies, any difficulty in finding them will be due to your handling. When going to look for a queen, open up the hive quietly, using as little smoke as you can—don't go blasting smoke about all over the place and thus get the bees on the run. Take out the first frame slowly and carefully. If it is a frame of stores I would put it down without further examination and go on to the next. This would be repeated until the first of the brood is reached. In the early part of the season this might be several frames from the start, whereas in the height of the active season the first frame should contain brood. With practice you will find you can get through the empty or store frames very quickly until the first frame of sealed brood turns up, when you can go back one to see if it contains unsealed brood. If it has then start your careful examination of the combs here and concentrate on seeing the queen and nothing else. The examination should be swift. My method is that as I remove one comb I look at the face of the next comb, and often the queen is seen at this time. Each comb is taken out and examined around the edges first, in case the queen is moving over the edge to the other side. Next turn your attention towards the centre of the comb, and take a quick look around the edge again in case she has just come through from the other side. Now turn the frame over and repeat the process on the other face of the comb. All this should be done quickly with no stopping to move bees about. In my experience, if you go through the chamber quickly, with as little disturbance as possible, you can usually see the queen easily, for she will still be on the frame where she was laying when you arrived. If you go through all the brood frames without finding her, then come back through again, examining each comb carefully and blowing on the bees to move them about as required, or separating clusters of bees with a finger. Careful searching in this way, keeping an eye on the floor and the wall of the hive as you remove or replace combs, will usually result in success. I would never go through a brood chamber more than three times at one manipulation as by this time the bees will be displaced from their normal coverage of the brood and will often be running madly about. In these conditions the queen is unlikely to be found.

If you have a queen that is small and hard to find—perhaps one you have failed to find a couple of times already by normal methods—the following will help. Get an extra brood chamber and stand it on a roof, or floor, behind the hive in which you wish to find the queen. Open the hive up and put the first pair of frames in the empty brood chamber, keeping them a couple of inches away from the wall. Take the next pair of frames and put them in with the first two, leaving a couple of inches between the two pairs. Now repeat with a third pair of frames. In the original brood chamber you now have five more frames and perhaps a dummy board. Space these out in pairs evenly across the box and leave them all for two or three minutes. The fact that light is shining on the two outside faces of each pair of frames means that the queen will move into the dark between one of the pairs. After a couple of minutes you should pick up each pair at a time and, opening the frames like the leaves of a book, you should be able to find the queen in one or other of them.

Another method is to 'sieve' the colony through a queen excluder. The colony is opened and the brood chamber lifted to one side; a new brood chamber is placed on the floor on the old site with a queen excluder between the floor and brood chamber. The swarm board is placed against the front of the hive and the bees are shaken from the combs on to this board. Combs are examined to ensure they are free of bees and that the queen is not still clinging on, and are then placed in the empty brood chamber. When all the bees have been shaken in front of the hive and the combs are in the brood chamber the crown board is put on and the whole thing left for half an hour. A few puffs of smoke every now and then will help drive the bees indoors. When they have gone in the brood chamber is lifted off and the queen should be found on the floor or on the underside of the excluder.

Methods like the last two should only be used as a last resort where a really useless queen must be found for removal. Often the problem is of a different nature: a vicious colony may need requeening and it is necessary to find the queen and remove her. The less experienced may find this a problem because it is difficult for them to concentrate on finding the queen when they have to spend most of their time controlling the bees. The best way to get over this is to pick the colony up and carry it away some distance from its usual site. The supers can be left behind to collect the flying bees who will return home. Thus most of the flying bees will be gone as soon as you get them on the wing and the queen can be looked for in comparative quiet.

There are various methods which make use of anaesthetic substances to quieten the bees. I do not recommend any of these as they have extremely bad effects upon the bees, and although the colony is quietened at the time it can be extremely vicious afterwards. I would therefore advise beginners to get the help of an experienced beekeeper to find and remove the queen from a vicious colony. I would advise the experienced beekeeper to put his armour on and to use all his skill in controlling the colony rather than trying to put them to sleep.

Looking for unmated virgin queens is very different from looking for the mated laying queen of a colony, and they are more difficult to spot and catch. The mated laying queen will normally be parading slowly around the comb with bees turning to her and attending her. The virgin, on the other hand, is likely to be rushing about all over the place, pushing bees out of the way and being snapped at by the bees. Alternatively, she may be quite still, burrowed into a lump of bees and concealed by them. It is usual, therefore, to look for a disturbance, or a trail of disturbance, on the comb and carefully to break up any clusters of bees to look for a queen. The virgin may rush across the comb flapping her wings, and may even take off and fly away. If this occurs, close the hive at once, leave it in peace and the queen will usually return, as only a queen that has already been on a flight and knows her way around the apiary will so readily take to the air.

Clipping the queen

Clipping queens is a very easy operation which causes no problem to the beekeeper and no pain to the queen. The main thing is not to be afraid of handling the queen, who is much more robust than many beekeepers imagine.

Some people can pick the queen up from the face of the comb, others find it necessary to make her walk up on to something before they can get hold of her. If you belong to the latter group hold your hive tool in front of her and pick her up as she walks up it. She can be shepherded along by forming a half circle around her with the fingers and thumb and allowing her only to walk forwards out of it.

The queen may be picked up by the wings or the thorax, but never by the abdomen. As soon as you pick her up, particularly when doing so by the wings, she will curl her tail around and sometimes she will sting one of her own back legs. To prevent this lift her straight on to the ball of the left thumb—this will allow her to clasp it with her legs and keep them out of danger. Remember a queen will *never* sting *you*, she is devoid of any aggressive instincts except against other queens. Having got her clasping the ball of the left thumb I allow her to move forward under the first and second fingers, which are held together against each other. Holding a queen in this way, pressure is applied from each side of the thorax and she does not usually struggle. Her wings remain folded and are both clipped together near the base of the abdomen. When clipping, one blade of a small pair of scissors is inserted under the wings and you are bound to touch her back. Wait before cutting! In

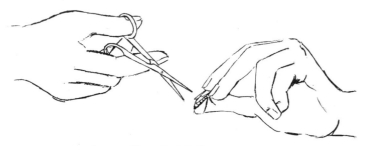

fig. 37 *My method of clipping the queen.*

many cases she will put a leg up to try and push the scissors off her back and will get it between the blades. If too rapid a cut is made you will have a clipped queen with five legs. Wait and cut carefully. I find this method of holding queens better than using just the index finger and thumb because the queen is much less likely to struggle and be dropped. If, however, you only wish to clip one wing you will have to use the latter method because it separates the wings. I would not advise the use of methods which trap a leg or two on one side and allow the queen to go round and round winding her leg up until she cannot move. There is enough trouble with paralysed legs without risking damage by these methods.

When putting the queen back into the colony she should always be placed on brood, and if possible released slowly so that she does not go rushing across the comb. Bees do not expect to see mother dashing about and may jump on her and sting before they realize who she is. If she runs or if the bees tend to chase her or snap at her as she passes, probably because they smell your scent on her rather than her own, immediately put the comb she is on back into the dark in the brood chamber. She will usually be quite all right, but whatever you do, do not try to defend her yourself or you will get her killed. The handling of queens is much more easily accomplished with yellow bees of Italian extraction. Black bees of the northern European race are very much more likely to kill their queen when she is handled.

Marking queens

Marking queens with various coloured spots or glueing paper discs or plastic caps on to the thorax are methods of recognizing individual queens, recording ages, etc. I can see very little use for the procedure except in experimental colonies when special information is required. The normal honey producing colony with a clipped queen is very unlikely ever to swop its clipped queen for another from some unknown source, so the fact that she has no wings will indicate it is the one on the record card.

However, for those who wish to mark their queens the following notes may be useful. If you use paint it can be any quick-drying lacquer or nail varnish which does not harm the queen. It is advisable to try out the effects of the paint on drones before using it on your queens. Acetone-based paints are safe to use. Paper discs can be made with the punch in an 'Eckhardt' marking outfit obtainable from the usual equipment suppliers. The paint spots or discs can be placed on the queen while she is held in one of the many types of cages sold for this purpose, or may be applied while she is held in the hand. My usual method is to pick the queen up as suggested for clipping. Once she is held between the two fingers and thumb the two fingers are separated slightly and slid down her sides, trapping at least two legs on each side and exposing her thorax. Marking is then easy.

Queen introduction

When introducing a new queen to a colony it must be done in such a way that both the colony and the queen are in the right condition to accept each other. The colony must be queenless, should not be in an excited condition from any cause, and should come in contact with the new queen fairly slowly. The queen should be in an undisturbed condition, should be hungry enough to solicit food from any worker who comes in contact with her, and if possible her odour, which will be that of a stranger, should be masked or her direct contact with the bees delayed until her scent has changed to something nearer their own.

Queen introduction during the early part of the season in April or May, and later in the year during late August and September, is easy, and queens usually can be introduced directly into the large colonies. But during the period between, when first swarming, and then the excitement of foraging or the frustration of being confined by bad weather when crops are in full bloom, make the bees more edgy, many queens will be lost if introduced into the colonies. Better results will be obtained by introducing the queen first to a nucleus and then introducing the whole nucleus to the full colony.

There are many methods of introducing the new queen to the workers, but I shall here only cover one method which should suit most people. This uses the introduction cage invented by Dr Colin Butler of Rothamsted, and known amongst beekeepers as the Butler cage. I choose this method because it is as satisfactory as any other, is by far the simplest and requires the least amount of equipment. The cage is made of wire gauze with about $\frac{1}{8}$ inch holes, formed into a square-sectioned tube $3\frac{1}{2}$ inches long and $\frac{3}{4} \times \frac{1}{2}$ inch in cross-section. One end is plugged with wood to about $\frac{1}{2}$ inch deep, as shown on page 158. I add a couple of long panel pins to this so that the cage can be placed between the combs, which are pierced by the pins thus preventing the cage from falling. Many beekeepers find that the large cylindrical hair curlers sold by large stores make very good substitutes

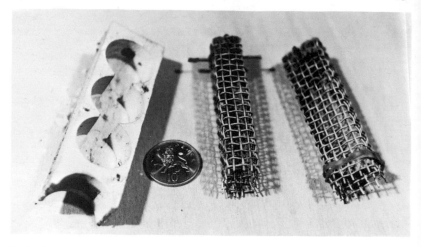

Left *An empty travelling cage. The farthest compartment is for candy. The far ends of the two Butler cages are blocked with wood, the open ends to be covered when the queen is inside.*

for Butler cages, the important requisite being the $\frac{1}{8}$ inch holes, so that the bees can feed the queen through them.

Queens should always be put into introduction cages on their own, never with their own accompanying workers. These workers may try to defend their queen against strangers and in the end get her killed. The queen is put into the Butler cage and is confined there by a single small piece of newspaper held in place over the open end by an elastic band. The cage should be hung in the broodnest in such a way that it has brood, preferably sealed brood, all around it. Escape from the cage then means that the queen is straightaway on brood, which is where bees expect to find her.

To introduce the queen to her new colony, the hive should be opened even more carefully than usual and the old queen found and removed. The new queen is clipped and run into the Butler cage, the piece of newspaper fixed on with the elastic band, and the cage then hung in the central part of the broodnest with the paper-covered exit roughly in the middle of the comb. The colony is then reassembled quietly and is left severly alone for at least six days. The mesh in the cage is open enough for the workers to lick and feed the queen, and get to know her. They will release her by biting through the newspaper within a few hours. After six days the colony can be examined to see if the queen is all right, and the empty cage removed.

During the swarming season, when there is excitement or robbing, the nucleus method of introduction is more likely to be successful. If the colony is one which is being requeened because the bees are trying to swarm, the old queen is killed at the time the nucleus is made up, and they are left queenless for at least a week or until the next inspection. If it is not a swarming colony, the failing queen may just as well be left to carry on as best she can until the new queen and her nucleus is ready for introduction.

In either case, however, the first thing to do is to find the old queen, and either kill her or place her in a matchbox with a few workers to look after her, according to the reason for requeening. A nucleus box big enough to hold about five frames is placed beside the hive, facing the same way. The broodnest is examined and a frame of emerging brood is found and placed, with the bees remaining on it, after a gentle shake to get rid of the old ones, in the nucleus hive. A second frame is put in the nucleus in the same way, but this should contain mainly stores. Three or four more frames are gently shaken first over the hive to dislodge the old bees and then into the nucleus box to dislodge the young ones, and these combs are then returned to the colony. A new, mated, laying queen is put in her Butler cage between the two frames in the nucleus and a dummy board is placed on each side. The nucleus and the main colony are both covered, roofed and left in position for about a week.

At the next inspection the nucleus hive is opened first. The queen is examined to see that she is all right, with no obvious infirmities, and that she has been laying for several days. The cage is removed, the frames moved to the centre of the box, and the dummies are taken out. The light will drive or keep the queen in the space between the two combs. The main colony is now opened and the queen is found and removed or, in the case of the swarming colony, the queen cells destroyed. A quick examination of the colony from which a queen has just been removed is advisable in case there are signs that the bees are starting to think of swarming or supersedure.

The combs are then pushed to one end of the brood chamber and the two nucleus frames lifted together and placed in the space from which they came some days before. It is a good plan to spray both the colony and the nucleus with water from a mist spray to stop the bees running about. As the spray hits the bees they close their wings tight over their backs and stand absolutely still for a few moments, after which they will start mopping up the water, which again helps in the successful introduction. When the queen has thus been introduced, the hive should be reassembled and not touched for at least six days. The nucleus method is a very successful one for use at any time.

Queen introduction is always more successful where new queens are being introduced to colonies of their own strain, and becomes more of a problem, with reducing success, as strain differences between the introduced queen and the colony increases. Most difficult is the introduction of a pure Italian yellow queen to a really black North European colony. The opposite is quite easy. It is suggested that Dr

Butler's discovery that the black race produces and requires more queen substance per bee for inhibition than the yellow race has something to do with this problem.

Making nuclei

No doubt many will buy their first colony of bees and then increase their number of colonies from this. Some may seek considerable increase, with the idea of running thirty or forty colonies. In all cases the same principles apply and must be considered before the nucleus is made. There are a number of uses for nuclei, such as mating nuclei (see Chapter 8) or requeening nuclei which has been mentioned already, and the same principles for making up apply to all.

Nuclei may be made up for use in the home apiary or for immediately moving away to another apiary. In the latter case it is easier, as one can make them to whatever strength one likes, confident that they will remain at that strength. When making them to keep in the same apiary they have to be made up extra strong in numbers, because the flying bees will return to their old site. It is, therefore, difficult to judge the number of bees that will stay with the nucleus, and it must be looked at each day for a few days to make sure that it is keeping its strength up. If too many bees have gone home then more must be put in from the same source. This is one reason why nuclei are best made with emerging brood. Brood at this stage in its life cycle is much less likely to suffer chilling and every bee that emerges is one that can help with the care of brood and will definitely remain with the new colony. The second reason for making nuclei with emerging brood is that the queen can quickly lay up the empty cells made available and because the emerged occupants of these cells will augment the number of nurses available for tending her brood. Thus the nucleus gains size quickly.

Nuclei can be made up one from one colony, several from one colony, or one from several colonies. The use to which they are going to be put, their size, and the circumstance of the unit in which they are being produced will have a bearing on how they are made, and the beekeeper will have to make up his own mind which method to adopt. Here I will give three methods which I hope will cover any eventuality.

Another variable is the origin of the queens which are to be used to head the nuclei. In some cases the beekeeper may decide to buy queens from amongst those available at the time. I would advise him to enquire amongst the experts, local or otherwise, to find out how these various strains have behaved in the past in his district. In this way it is often possible to avoid disappointment later. I hope the beekeeper will be encouraged to practice queen rearing himself, as described in Chapter 8, but one way of getting queens must be mentioned and condemned. This is by making up a small four or five-comb nucleus and allowing it to make its own queen on the emergency principle. This is the most certain way of getting a 'scrub' or useless queen. To ask a small colony which has just been made up, and is therefore far from balanced with bees of all ages, to start from scratch and feed a queen so that she reaches optimum condition is asking the impossible and, although I know this is a method still advocated by some beekeepers, I cannot condemn the practice too strongly.

The basic technique of making a nucleus is as follows. The colony from which the nucleus is to be made is opened up quietly, the queen found and placed in a match box in a safe place with three or four workers. The brood combs are examined and those containing emerging broods are selected and placed in the nucleus box to the number required, i.e. four for a four-comb nucleus. If the nucleus is to be taken away immediately to another apiary all the bees should be left on the frames of brood and another two combs of bees shaken into the nucleus box, which should then be immediately shut up ready to take away. The nucleus is taken to the new site and opened up, allowing the bees to fly. It can then be gently smoked and its contents transferred to a full-sized hive. It can of course be made up in a full-sized hive from the start, if this is convenient. Once transferred, a new laying queen should be introduced with a Butler cage (see page 158), and the small colony completed with four frames of drawn comb or foundation, and fed a gallon of syrup. It must then be left alone for at least six days, after which it can be built up.

If the nucleus is to stay in the home apiary the frames containing brood should be slightly shaken over the hive to remove old bees before being put in the nucleus box, and then four more frames should be shaken lightly over the brood chamber and the rest of the bees shaken into the nucleus box or new full-sized hive. The nucleus with four frames of brood is then put on its permanent site, the entrance lightly blocked with grass, the queen introduced and the extra combs added as above. The main difference in making this home-apiary nucleus is the effort made not to include old bees, the extra bees put in to allow for some to return home, and the fact that it is not fed for six days. The reason for this is that if a feed is put on right away, the old bees may carry the message back to the rest of the colony and the nucleus may be robbed out. When making such a nucleus, therefore, it is necessary to ensure that it has at least six days' supply of stores in the combs. Syrup cannot be given it until after six days, when any old bees have gone home and the queen will be established. The grass stuffed in the entrance will help to delay the bees flying and will impress on them the fact that they are in a new place. This may prevent some from returning to their old home.

In any nucleus made for your own use you have control of the feeding and management. If, however, you are making nuclei for sale they should be made with at least one frame of stores, as you have no idea what treatment they will receive when they leave you. You must make them self-supporting from the start. Also, the queen in the nucleus must be established, and therefore it is best to pack the reigning queen off with the nucleus and place your new queen with the remainder of the colony.

Any colony which has lost a nucleus will lose some of its possible honey production, but how much will depend upon the initial size of the colony, the time of year and the availability of nectar supplies at the time of, and just after, the making up of the nucleus.

If several colonies are available and the amount of increase required small, a nucleus can be drawn from several colonies. This can be done by taking one comb of brood from each of four different colonies *without any bees*, and putting in all the bees required from another colony. In this way the honey production of the colonies providing the nucleus will hardly be affected.

Nuclei of the type dealt with so far should show at least some crop return in the same year providing the weather is good and forage is there to provide nectar. If increase is required and no honey crop is expected from them the same season then much smaller nuclei can be made. This means that the colonies from which they are made suffer less reduction, or that more nuclei can be made from the same number of original stocks. Nuclei made up in June with one frame of brood and one of stores, with bees to cover, and given a new laying queen right away can be built up to full-sized colonies by winter in the vast majority of years. Swarming colonies can be broken up into nuclei of this size, each with a good queen cell, and although these are slower to build up because of the delay of mating before any laying will commence (the population not increasing in size for about one month), after that they can be built up quite quickly and in most years will be of adequate size for good wintering.

Uniting colonies, nuclei and swarms

The opposite of making increase is uniting colonies—cutting the numbers of your colonies down by joining some together. Usually uniting is done in autumn or spring: in autumn because some of the colonies are too small for wintering well or the beekeeper wants to cut the number down; in spring because a colony has come through the winter queenless, or with a drone-breeder queen, and no mated laying queen is available to take the colony on.

As flying bees will return to their old site it is necessary to bring colonies to be united to within 3 feet of each other. To achieve this they can be moved up to 3 feet per day. If your colonies are already arranged in pairs, one from each pair can be united from any distance, as the flying bees will return to the remaining colony of the pair.

Full-sized colonies are best united by the *paper method*. If both colonies have queens, find and kill the poorer one of the two and leave its colony to settle down. In the evening go to the other colony, open it up as quietly as possible by removing the crown board. Place a sheet of newspaper over the top of the frames, prick a few holes in the paper with a pin or the corner of your hive tool. Hold the sheet of newspaper down with a queen excluder. Then quietly lift the queenless colony on to the top of the paper and excluder and leave alone for a minimum of six days. By the next morning the colony scents will have mingled and the bees will be united and most of the newspaper will be in a heap of small pieces below the entrance. The queen excluder is only needed to hold the paper down—in my experience there is always a wind blowing when you want to unite. At the next inspection clean out the remains of the newspaper and, if there were supers on the bottom hive, place the brood chambers directly above one another, leaving the queen excluder between them. In twenty-one days all the brood in the top brood chamber will have emerged and this can now be removed, bringing the hive back to normal.

Nuclei can be united by the paper method, but when they are in full-sized brood chambers which are not full of frames they are usually united directly, while in some way masking the scent of the bees and giving them plenty of work to do. This can be done by dusting them with flour, spraying them with water, or syrup, preferably scented. Usually the two boxes are brought close together, the frames taken out one at a time, with the bees on, sprayed or dusted with flour on both sides and placed in their new full-sized hive. The nucleus with the queens is done first and then the queenless bees are treated in the same way. Bees remaining in the nucleus boxes are dusted or sprayed and then poured over the top of the frames. This is about the only instance in beekeeping where a bit of bumping about and heavy handed treatment will aid the work rather than hinder it.

Swarms may usually be united at the time of hiving by throwing them down one on top of the other, when they will all sort themselves out and the two queens will be reduced to one. Swarms that have been hived for up to a week will generally accept other swarms hived in the traditional way into their midst. Usually there is no fighting at all, and never in my experience more than a few minor skirmishes.

Finally a word of general warning regarding uniting. Uniting two poor colonies does not make a good one. If they are poor because the queens are poor, or because they are diseased, then the resulting unit will still be headed by a poor dud queen or will be diseased. It is best to find out why colonies are poor and to deal with their problem before, rather than after, uniting. It is best to unite poor colonies, providing they are free from disease, with good average colonies who can make use of the extra population and provide some net gain.

8 Queen rearing

We have stressed the need to have young mated queens available at various times during the active season: in the spring to replace a 'poor queen' and in the swarming season to replace queens in colonies that have made up their minds to swarm. Also at those times all through the active season from April to the end of July when queens may suddenly fail and need replacing, and at the end of the season when two-year-old queens should be replaced with young ones. In other words, for many reasons the useful length of life of queens may vary considerably and some preparation must be made to provide replacements which are of good quality and breeding.

With the honeybee there is a more obvious difference between the concepts of quality and breeding than with many other animals. The quality of a good queen with excellent inheritance can be heavily concealed by poor nurture during her larval development. Bee breeding is difficult and although extremely interesting may have to be left to the beekeeper with a large number of colonies. Queen rearing, on the other hand, can and should be practised by all beekeepers. The queens that are to be used in the apiary should be the product of thought and planning. They should not be the queens that the colony happens to make, when it can no longer hold together with the queen it has.

We know that the fertilized egg of the honeybee can be turned into either a worker or a queen dependent upon how it is housed and fed. We also know from experience and research that the best queens are those produced in large colonies where there are lots of young bees and plenty of pollen for them to feed on when they are making 'bee milk'. The queen larvae are then fed to a maximum and grow large, and with a large number of egg tubes in their ovaries. In contrast, the small nucleus will never be able to produce a top-line queen. The nucleus is usually struggling to build up and has as many worker larvae mouths to

Queens in travelling cages being recorded
as part of a large-scale breeding programme.

feed as it can manage. To expect a number of queen cells as well is to ask for a poorly-fed queen.

The best queens are produced from very young larvae, or eggs. Research shows that the larva which is treated as a queen from the start produces the heaviest queens with the highest number of egg tubes and the largest spermatheca. It also shows that the reduction in these factors occurs as progressively older larvae are taken for queen rearing. Thus if we are to produce queens for use in our own apiary we should produce them in as large a colony as possible: one in which there are plenty of young bees to act as nurses, and ample pollen. Finally, the worker larvae from which the queens are to be made should be as young as possible when they are started off on their careers as queens.

The process of queen rearing can be broken down into four separate parts: the provision of the colony which is to produce the queens, usually called the *cell-building colony*; the selection of a colony which is to provide the larvae—this is the *breeder colony* and it contains the breeder queen; the process of giving the larvae from the breeder queen to the cell-building colony, and finally the removal of the ripe queen cells from the cell-building colony before the first virgin queen hatches (or she will kill all the rest) and the placing of these queen cells into small 'mating nuclei' from which they can fly, mate, and in which they can start laying.

Let us look first of all at a selection of breeding stock. It would be stupid not to take advantage of the process of queen production to increase the value of our stock as much as possible. Bad characteristics which can easily be recognized can be bred out very quickly, and these include stinging, following and excessive running about on the comb when being manipulated, all separately inherited and tiresome. Running about on the comb can be so bad that when combs are lifted from the hive the bees on them run down to the bottom of the combs, form clusters and drop off. It requires little imagination to picture the problem if this is happening when you are looking for the queen. These characteristics should be culled from your strain of bee as quickly as possible by avoiding producing queens from colonies which show them, and by replacing the queen in such colonies as soon as possible. The sooner they are gone the better, because all the while they are there they will be producing drones which may mate with the young queens and pass the bad traits on to future generations.

Persistent swarming is another inherited trait that can be reduced by culling—that is by replacing those queens whose colonies show it. Swarming is the bees' natural method of increasing the number of colonies, or the number of sexual females, whichever way you wish to look at it. Without swarming reproduction does not take place, and from the point of view of the species as a whole this would reduce its ability to withstand adverse conditions. I therefore feel that it is not possible to envisage a useful bee from which the swarming instinct has been entirely eliminated. It can, however, be greatly reduced, and for this reason I would try to breed from bees which neither try to swarm every season, nor make large numbers of queen cells when they do. I would breed from colonies that once having made up their mind build up to nine or ten cells, but colonies such as one I had in Devon which produced 153 queens and queen cells at one time should be culled as rapidly as possible.

Breeding for honey production is much more difficult because its characteristics cannot be assessed in any meaningful way. Individual colonies which produce very large surpluses of honey may do so for many reasons other than the inheritance of a very high work rate. They may just be very good robbers, and have stolen their honey from other colonies. They may be in a position in the apiary where a lot of bees drift in on a prevailing wind. They may always be that truly exceptional case which has inherited genes which all add together to give a very high production, but this is a fortuitous happening which is not possible to repeat in the offspring. The only useful method is to look at the family from which a queen comes before she is chosen as a breeder. Her sisters should all be equally good and all their colonies acceptable to the beekeeper.

You often hear it said that you should not breed from the exceptional colony. But often this is then altered to 'you should not breed from your best colony', which is not necessarily correct. If you have only three or four colonies, or even a dozen, you are unlikely to have an 'exceptional' colony in your apiary. These are by definition very rare and the chances of their turning up amongst a few hives is very small. The beekeeper with just a few hives is best advised to breed from his best colony. He may come unstuck once or twice in a lifetime but this is a chance worth taking. If he has a large number of colonies then he is best advised to breed from a queen belonging to a good family. The beekeeper with the small number of hives can of course band together with a number of other beekeepers and by selecting over all their colonies practise 'family selection'—the result will be much more successful in the long run than working alone. So much for breeder colony selection.

The provision of the cell-building colony will depend very much on the number of hives you are catering for and the number of queens you wish to produce. I will therefore deal with the subject at three different levels. One for the beekeeper with up to about ten colonies, secondly for the man with up to fifty colonies, and finally large-scale rearers.

The small scale beekeeper will do best by deciding to work his best colony on two brood chambers. This is the colony that is building up most rapidly in the early season. If this colony is given a second brood chamber of drawn combs the bees should spread up into it very rapidly. If the colony which is to be given the second brood chamber has an arch of honey in the top of the frames of the original brood chamber, get them to shift this by scraping the capping with the hook of your hive tool, thus laying bare the honey. The bees will usually then remove the honey and take it to the top box. The colony should be built up rapidly to as large a size as possible by the third week of May. This date will depend upon the time at which you can safely start queen rearing and expect to get the queens mated.

Once the colony is using most of the brood chambers add supers as required: by the third week of May it should have at least one super, if not two. The colony is now ready to produce queen cells, and you should act as follows. First find the queen and place her in a match box with a few bees to look after her. The two brood chambers can then be sorted through. Put all the *unsealed* brood in one and make the box up with sealed brood and one good frame of pollen if this available. Put the sealed brood on the flanks—i.e. nearest the hive walls—the unsealed brood between them and the pollen comb in the centre. The remainder of the combs are put in the other brood chamber, the queen freed on to one of them, and the brood chamber then replaced on the floor on its original site. The supers are put on next above a queen excluder, and if there is only one super at this time I would add a second. A second queen excluder should now be placed above the supers and the brood chamber containing the young brood placed on top covered by the usual crown board and roof.

The young nurse bees will be drawn to the top brood chamber by the presence of the unsealed brood, but the fact that they are isolated from the queen by two excluders and two supers and that the full transfer of food, and hence queen substance, will not take place between the bottom and top brood chamber bees, means they will usually make a small number of queen cells. In this case you are using this colony as the breeder colony as well as for the production of cells. It may be, however, that the queen you would like to breed from is not capable of building up a colony large enough for the above procedure. In this case you must insert a marked frame of eggs and very young brood taken from the colony from which you wish to breed your queens, and remove any queen cells produced from their own brood by the bees in the queen-rearing colony. If you are lucky with this method and get a satisfactory number of cells, as soon as these are ripe—about ten days after the colony is split as described—you may either cut them out and distribute them to mating nuclei, or the top brood chamber combs and bees can themselves be split up into mating nuclei using some of the queens. You will only be able to split it into about three nuclei if you are going to have sufficient bees in each.

The second method is for the production of from twenty to forty queens and is much more positive than the above method. In the early season the technique is the same—a very large colony is built up on two brood chambers. Because the beekeeper has a greater number of colonies he can take frames of brood from colonies which are building up well and give these to the cell-building colony, thus building it up to a massive size. When the time for queen rearing comes the queen is found and removed on to a small two-frame nucleus. The rest of the colony is made up with a super at the bottom, on the floor, then a queen excluder with a brood chamber above it. The brood chamber should be filled with eight frames of sealed brood, one frame of unsealed brood and one frame of pollen, the sealed brood and pollen being placed in the centre of the brood chamber. Assuming an eleven frame chamber this leaves an empty place which can be filled with a dummy board. The colony should now be given all the bees from the rest of the combs, both brood and super. This is done by shaking them into the brood chamber just set up, and the colony may then be given a feeder of syrup and closed down. Any surplus brood and supers should be dispersed amongst the other colonies—as these frames have no bees they cannot be given to their own queen as she has not enough workers.

The main colony, now congested to overflowing and queenless, will make queen cells. I would leave the colony for two or three days and then remove the new queen cells in it, shaking the bees off the combs so that none is missed, and give them a frame of larvae which are to be turned into the queens we want from the breeder queen. These larvae should be put in the centre of the broodnest between the frame of young brood and the frame of pollen. Queen cells will be constructed on this frame and will be sealed in four days, so a second batch of larvae from the breeder queen can be given at this time if required. Cells will be ready to be distributed ten days after the larvae are put in, so if two batches are required the colony will be cell building for 17–18 days from the time the queen was removed. As soon as their role of cell-building colony is completed, the original queen in her nucleus can be put back and the colony brought back into honey production.

The final method is for the beekeeper who requires a considerable number of queens. It is very like the last method but involves combining two large colonies in a special brood chamber that takes 13–15 frames. This massive colony is kept going from about the third week in May to the end of July, and larvae for queen rearing are placed in every three or four days. As the worker brood hatches, more is added from other colonies and this prevents a fall in population. Such a method can produce several hundred queens in the course of a summer but requires a back-up of mating nuclei available in the required quantity.

Having set up this cell-building colony, two or three days later the whole colony is looked through and, as with method 2, any queen cells are destroyed, and larvae from the breeder colony inserted.

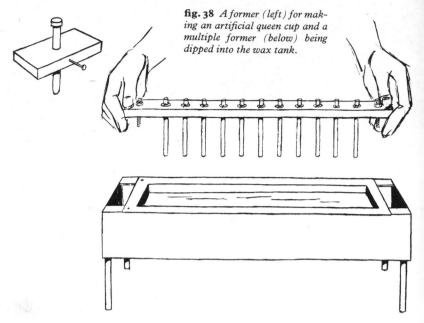

fig. 38 *A former (left) for making an artificial queen cup and a multiple former (below) being dipped into the wax tank.*

Grafting

Insertion of larvae can be done in many ways but my own preference is for the 'Doolittle' method or, as it is commonly termed, 'grafting'. This is the process whereby a number of small waxen cups are made by the beekeeper. These are attached to bars constructed in the usual size frame. Small larvae are then transferred from their comb in the breeder colony, one into each cup. These are then placed in the cell-building colony for the queenless bees to turn into queens. I find this method the easiest, quickest, least messy and most reliable of all the methods generally used. This is how it is done.

The wax cups are prepared by dipping a wooden or glass former into molten beeswax. The former can be made from $\frac{1}{4}$ inch dowelling or glass tubing, the ends of which are rounded off and well smoothed. If a lot of cells are to be made a bar with a number of formers can be dipped, giving several cups for each dipping, as drawn above. The formers will stick to the wax unless they are wet, and for this reason they are placed in water several minutes before use and are dipped in again between each application. Wax is melted in a small water bath—nothing more elaborate than a small empty meat or fruit tin standing in an old saucepan is needed, though a special double-jacketed trough as shown in fig. 38 can be used. The wax should be good, clean wax and should not be heated much above melting point. The former is then removed from the water, shaken to get rid of excess

fig. 39 *Artificial queen cups being fastened to a bar with wax. The bar can then be swivelled round so the cups hang vertically.*

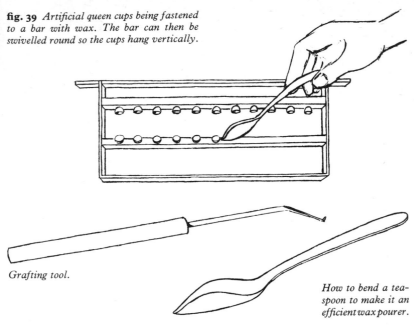

Grafting tool.

How to bend a tea-spoon to make it an efficient wax pourer.

water and dipped about five times into the molten wax to a depth of about $\frac{5}{16}$ inch. Some beekeepers try to dip progressively less deeply each time to provide a thin edge to the 'cell'. I have never found that this helps in any way and usually dip to the same level using a depth guide on the former. The single former is dipped until the wooden crosspiece hits the sides of the container and the multiple former has the two bolts at the end which can be adjusted for height and which contact the sides of the trough. After dipping, the former is placed back in the water for a few seconds, when the cups can easily be twisted off the ends.

The cells are fastened to bars on a frame made up as illustrated above. The bars can be fastened to the frames in several ways but the main thing is that they must be easy to get at when fastening the cells on. Equally, the cells may be put on in several ways, and here the main thing is that they must be easily removable for taking away to mating nuclei, and should be robust enough for easy handling. The cups can be fastened to small squares of wood, which are then stuck to the bars, or by melting beeswax on to the wooden bars to about $\frac{1}{8}$ inch depth before they are stuck on, illustrated overleaf. In both cases they are stuck on to the bars with molten beeswax. A useful tool for both building up wax on the bars and then fastening cells to it is a teaspoon bent in a vice as shown.

Left *Grafting a larva into an artificial queen cup containing diluted royal jelly.* **Right** *The size of the larva on the end of the grafting tool in relation to a pin's head.*

When the cells have been fastened to the bars they are ready for use. The next thing to do is to prime them with a small quantity of dilute royal jelly; this can be obtained from the queen cells which have to be destroyed in the cell-building colony. The royal jelly should be diluted until it will just drop from a matchstick or something of about that size. An eye dropper can be used instead of a match to place the jelly in the cells, but this should only be done a few minutes before the larvae are to be transferred as it will soon dry up. The use of jelly is not absolutely necessary but it is considerably easier to float the larva on to a drop of liquid than to get it off on to the dry floor of a cell. If possible the whole apparatus—frames, bars, cells and jelly—should be kept warm at just over blood heat so that the larvae are not chilled. It is also advantageous to have high humidity, and this can be easily obtained by boiling water close on hand. Good queens can be produced in less than ideal conditions as long as reasonable care is taken to prevent chilling.

Having got this far, the beekeeper will go to the breeder colony and select a comb containing very young, just-hatched larvae. The bees are brushed, not shaken, from the comb, as heavy shaking may displace larvae and make them more difficult to pick out. The frame should be covered with a cloth and taken to the apiary shed, or wherever the grafting is to be done, as quickly as possible. The artificial cells are primed with royal jelly and the selected larvae are transferred, one to each cell, with a transferring or grafting tool, shown in fig. 39, which can be made of stainless steel, plastic or wood. The larva should be approached from the back (shown above) and the flat end of the tool slipped under it so that it can be lifted. Transfer to the liquid in the cell is easy: it is only necessary to pass the end of the grafting tool through the drop and the little larvae will be floated off. The larvae should be under thirty-six hours old and as small as possible—about *half* the size of a lower case letter 'c' in this book. A good light is necessary to see them and this will help to provide heat as well. The novice may find it is easier to find a row of larvae, and then to run a warm sharp knife along their cells, levering the bottom half back to an angle of 45° so that the larvae are easier to get at and easier to see.

As soon as the larvae have been grafted, check to see that you have one in each cell, and covering the new frame with a cloth carry it to the cell-building stock and place it between the unsealed brood and pollen without delay. If a space has been left open for the graft frame, bees will have clustered in this space and will transfer directly to the graft as it is slowly lowered into place. If the bees do their job, these cells will be ready for dispersal to the mating nuclei ten days after grafting.

Mating nuclei should be prepared eight days after the cells are grafted so that they are queenless for two days before the grafted cells are put in. This puts the nuclei into the right condition to receive cells. The cells themselves will not be ready to hatch for two more days, so at

A grafting frame with cells ready for distribution to mating nuclei. In the lower row the fifth cell from the left is too small for use and the end cell has been rejected by the bees.

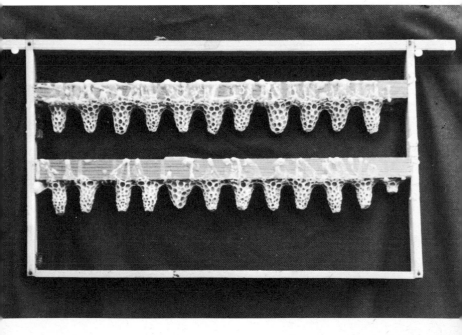

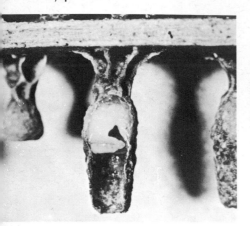 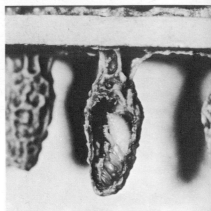

Grafted cells opened to show the contents. The larva (left) is almost fully fed, of good size and well supplied with royal jelly. The pupa (right) is well developed and large.

the time of dispersal from the cell-building colony to the mating nuclei the queens in their cells are still quiescent, wrapped in their pupal skin. This is the time when cells are most readily accepted by the mating nuclei, and unlikely to be damaged by the workers. The use of queen-cell protectors, recommend in some of the older literature, is unnecessary.

Mating nuclei

It is the provision of bees for mating nuclei which has prevented many beekeepers from rearing queens. They dislike breaking up colonies for this purpose when they could be producing more honey. It is true that you reduce honey from these colonies as well as the ones you use for cell building. The loss will, however, be more than compensated for by the use thereafter of good, well-nurtured queens in all your colonies. Nor do you need to take a great deal from your colonies to make the mating nuclei, as these can simply consist of one frame of brood and one of stores.

Making nuclei is dealt with in Chapter 7 and the same rules and problems arise whether the nuclei are to be used for mating or increase. I should like to deal here with what I consider to be the most economical method of using mating nuclei, which also covers the over-wintering of some young queens for use in spring and early summer. Another advantage of this system is that the nuclei are permanent and, unless they are lost in a very bad winter, need making only the once. The method is based upon double nucleus boxes as shown in fig. 40.

The double box uses the same frames as in the rest of the apiary: it is therefore a normal brood chamber into which, halfway along each side

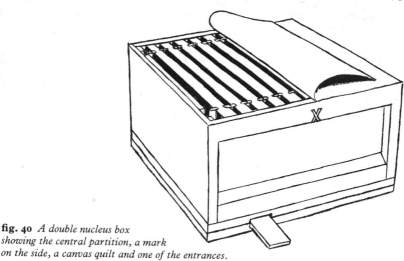

fig. 40 *A double nucleus box showing the central partition, a mark on the side, a canvas quilt and one of the entrances.*

wall, is worked a groove which takes a central partition. This partition must be fitted to the hive so that it is absolutely bee-proof between the two nuclei. If they can contact each other, and particularly if the queens can get at each other, only one queen per box will be the result. To ensure they are bee-proof a floor must be permanently and securely nailed on to the brood chamber. The central division can then be fitted to the floor and marked so that it is always put in the same way. I would suggest three marks to save problems. These are placed one on the division board, for example on the top right hand corner, with a similar mark on the side wall directly beside to locate the partition in the right hive and the right way round. If the partition is out and the brood box being used as a normal hive full of bees it is helpful to know which board belongs to which hive before going to the colonies. It is best therefore to repeat the mark a third time on the outside of the side wall where it can be easily seen. When the partitions are in, a canvas quilt is more convenient than a wooden crown board for each half. Fig. 40 shows the canvas quilt nailed to the top edge of the partition—handy when opening the nucleus with one hand and carrying queen cells in the other.

The nuclei will be made up on each side of the partition with a frame of sealed brood next to the partition followed by a frame of stores and a dummy or a frame feeder, as shown in fig. 41. Sufficient bees are put in to keep the whole lot warm and, if the nuclei are being made from colonies in the same apiary, to allow for the return to their old home of any old bees. It is a good idea to check after one day that the nucleus strength is adequate. Two days later the ripe queen cell will be put into the nucleus after the beekeeper has shaken the little colony through and removed any of their queen cells. The queen should be mated in 10–14 days if the weather is fine and from then on the little colony will build up very rapidly. They will need more comb very soon and I would give them foundation only as this will be pulled out beautifully and will also hold them back somewhat from too rapid a build-up.

The number of these boxes required and the way they will be used will depend upon the number of colonies kept by the beekeeper. One double box in which two queens can be over-wintered should be enough for up to about eight colonies. More queens will of course be needed for summer requeening, but these could be mated in single nucleus boxes, as on p. 78, the bees of which would be dispersed when the queen is moved. In fact the entire nucleus could be used to requeen any colony in need. The double box would be given its cells and then built up to five frames on each side which, if well fed, should be fit to overwinter. In good seasons these nuclei with new young queens build up very quickly and it may be necessary to take sealed brood away to prevent the population overflowing. The sealed brood can be given to one of your other colonies and the bees it produces can work for you to get honey.

For larger beekeeping enterprises the number of double boxes will be increased pro rata, using a ratio of 1:8 to calculate the number of double boxes needed. This ratio allows the large beekeeping enterprise using the continuous cell-building colony to have all the mating necessary done from these double boxes. The method here is to remove the new queens as soon as they are laying, or as soon as their first worker brood is capped and two days later to put another queen cell in from the cell-building colony. In this way about three or four queens can be mated in each nucleus during the season.

Knowing when the new queen will emerge in each nucleus, I would look for her on that day: she is usually easy to find because she is light in colour and fairly slow in her movements, not having had time to 'harden off'. A sight of her will confirm if she is in good order, with all the necessary legs and wings. The workers will tolerate the examination when the queen has *just* emerged but after this they

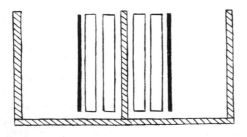

fig. 41 *A double nucleus box with two frames and a dummy board on each side of the central partition.*

Brother Adam's queen mating apiary on Dartmoor, using quadruple nucleus boxes. The single-pole stand is ideal for mating nuclei because it gives easy access on all sides.

become progressively more likely to kill her if the colony is disturbed. Thus, look on the day of emergence or the following day, but thereafter leave them alone for not less than ten days, in which time, if the weather has been reasonable, she should have been mated and possibly have started to lay. The tolerance of examination varies with the strain of bee and you may find that you can look at your bees again in under ten days, but I would say not less than six days. By looking at the nucleus on the projected day of emergence, the cell can be examined and the queen 'pulled' if she has not emerged, or the contents discarded if she is anything other than a good queen.

A pulled queen will be quite all right providing she can stagger along until she dries out and hardens. Any cell which is discarded, or any one from which the queen has emerged and vanished, should be replaced immediately with a queen cell from the cell-building colony. This saves time and prevents these mating nuclei from becoming truly queenless.

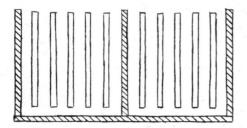

fig. 42 *A double nucleus box with five frames on each side of the central division.*

The beekeepers who do not need to use continuous cell-building colonies (impractical unless they need about a hundred queens a season) will probably find it most economical to scale up from the small-scale method. That is, to do their queen rearing in three or four successive grafts, making up nuclei to accommodate these; some in double boxes and some in single boxes.

Double boxes are used to overwinter the last two queens mated in them. To do this they should be built up on to five frames on each side of the division, and each side fed down as heavily as possible with separate feeders in August. The more bees they have, the better they will survive the winter. If several queens are being mated in them during the course of the season, with the frequent manipulations this will entail, it is best to feed Fumidil 'B' in the autumn syrup (see page 132) at least every other year. This prevents the build up of nosema in the mating apiary (see Chapter 9).

When the next active season arrives one queen can be used from each double box to correct weaknesses in the honey-production colonies. The nucleus is opened, the queen removed into a travelling cage (see page 164) with about twelve workers put in to look after her, and in this she will be all right for a week or so, although she should be used as quickly as possible. The easiest way to put bees into the cage with her is to take a comb out which has honey in it on which the bees are feeding at the time. When they have got their backs to you and their heads in the cell they cannot see your fingers arriving and usually their wings are neatly folded ready for picking up. If picked up by both wings they cannot curl their tails around to sting, and can be easily introduced into the entrance of the cage. The queen can be clipped before she is put into the cage which makes her removal from it later so much easier. For food a little candy should be put in the cage. 'Queen cage candy', made by mixing ground fondant or icing sugar with honey, is best. It should be made as stiff as possible, but this particular candy has the advantage that it never goes rock hard, as does candy made with water.

After the queen is removed from one side of the double box, all the bees can be dusted with plain flour, using a flour dredger, while they are on the combs and the central partition removed. By the time the two little colonies have cleaned up the mess of flour they will have become peacefully united. This colony can then be built up as a unit until the queen rearing starts again when it will be broken up to provide the two mating nuclei, each consisting of one frame of brood and one of stores plus bees to cover on each side of the replaced central division.

The colony is likely to have built up by this time so that there will be a considerable residue of bees and brood and, of course, a surplus queen. This surplus portion can be used in several ways. It should normally make a nice five-frame nucleus, and this can be used to make increase or to replace a colony lost in the winter; it might be added to a weak colony from which the queen has to be removed, or it could be sold. Alternatively, it can be split up into four or five mating nuclei if these are required. This method avoids the continuous need to break up production colonies to provide mating nuclei. In effect, the queen rearing becomes separated from the honey-production side, with advantages to both.

The whole process of queen rearing should be carefully recorded in an apiary book, giving dates of making up cell-building stock, recording each graft, with the number of cells grafted, the number accepted and the number put out to mating nuclei. Each mating nucleus should be recorded as to when it is made up, when queen cells are introduced, whether virgin queen is seen, date when mated, and date removed. In this way a complete picture of your queen rearing can be seen, the origin and history of every queen known, and mistakes noted and analysed so that your technique is improved over the years. It is a great pity that more beekeepers do not practise queen rearing enjoying the great interest and excitement which goes with it, and the personal satisfaction of seeing a large laying queen which you have been instrumental in producing from the time she was a small larva.

9 Pests and diseases

This chapter deals with the problems of disability, disease, poisoning, pests, etc. Some of these are bound to turn up at some time or other if bees are kept for any length of time. Do not be down-hearted; neither you nor your bees have been singled out by fate to suffer this catastrophe. It is a normal happening in the life of any animal or plant and as a beekeeper you deal with it and that is an end. Nor is there any stigma in having disease turn up in your colonies. The secrecy which seems to surround outbreaks of disease is ridiculous. If we all talk to one another about the disease in our colonies and we shall find it is not of very high incidence and could be less if dealt with promptly.

Queenlessness

There is considerable misunderstanding in the minds of many beekeepers on this subject. Several times in every year beekeepers tell me they want to get hold of a queen because they have a queenless colony. When asked, 'How do you know it is queenless?' the reply is invariably, 'Because there is no brood.' Although it is true to say that brood is usually absent from queenless colonies the converse is far from the truth. A colony with no brood can have a perfectly good young queen who has not as yet started to lay. In my experience the number of colonies that become queenless by natural means is very, very small. Usually the queenless colony, particularly during the main part of the active season, has been made so by some mistake by the beekeeper.

Recognition of queenlessness is far from easy if one is just relying on conclusions drawn during examination of the colony. The main signs are that the colony is more irritable than usual, the bees seem to be less well-organized on the combs, very few brood cells will be polished up ready for the queen to lay in—certainly not a large circular area of such

A lesser wax moth on the silken web of destruction created by its larvae as they devour the honeycomb.

cells. Pollen in the broodnest will be shiny from being covered with honey to prevent it going mouldy whilst it is not being used. Often there will be some cells with little hoods drawn out from the top walls and often these are covering pollen, and in some cases an egg from a laying worker. All these signs are straws in the wind pointing towards queenlessness but none is conclusive.

Most of the year, however, there is one sure way of finding out whether a colony is queenless or not, and that is by putting in a 'test comb'. Another colony is opened up and a frame of very young larvae and eggs is taken out, all the bees shaken off, the combs pushed up together again and an empty drawn comb put in at the side. (This empty frame could come from the 'queenless' colony.) The frame of brood is then placed in the centre of the 'queenless' colony's brood chamber. If the colony is queenless they will make queen cells on the brood which can easily be seen four or five days later. If they have not made queen cells they have a queen of some sort and the next job is to find her. The queen could be an old one who has given up laying or a young one who has not yet started. Quite often the frame of brood will have provided the colony with a focus and the queen will be found on this comb. The usual procedure, therefore, is to open the colony, go straight to the test comb and remove it. If there are no queen cells then it is examined for the presence of a queen. Once the position has been clarified the remedy is obvious: if the old queen is present remove her and requeen with a mated laying queen, or if there is a young queen present leave her to start laying.

The only time the method of using a test comb breaks down is directly after a colony has swarmed. At this time even with a virgin in the hive bees will often make queen cells on the test comb to try to carry on swarming. Once a colony has swarmed, however, I would never think of it as being queenless until at least a month after the last swarm left, after which time a test comb will usually give the true position.

Drone-breeder queens and laying workers

The presence of these two pests is very easily recognized; in the former case by the large drone cappings raised on worker cells (see opposite), and by the presence of half-sized dwarf drones running around the brood area in the latter. It is not so easy for the beginner to differentiate between the two causes, and even the experienced can come to the wrong decision. The drone-breeding queen is usually quite obvious when she first starts to produce drone brood in worker cells because these will be mixed in with ordinary worker brood. As time proceeds the amount of worker cappings reduces and the number of drone cappings increases. Whilst there are *some* worker cappings left it is obviously a queen laying, and not workers, but as the queen gets progressively shorter of sperms so the time will come when nothing

A drone-laying queen creates this ragged and distinctive pattern on the brood when the workers try to alter the cells to accommodate the larvae. Laying worker cells are similar, but usually in scattered patches. If some larvae are dying this state of brood can be confused with AFB (see page 197).

but drone cappings is present. The colony will be still reasonably large—with at least two or three combs of brood—and normally the beekeeper who regularly examines his colonies will see what is happening and will have solved the problem easily by requeening.

The real difficulty can be when first examinations are made in spring. Here you may find a small colony with brood only on one or two combs and all of it capped drone. Is this a drone-laying queen or laying worker? A queen will still be laying her eggs in an orderly manner, and the actual area of brood will be fairly solid with few empty cells. Laying workers, on the other hand, lay in a haphazard way, with bunches of cells here and there and not an oval or discrete solid area. Usually with laying workers there are also endeavours to build and produce charged queen cells, and while this may also occur with a drone-breeding queen, it is unusual.

If you feel that the broodnest is tidy enough for a queen to be present you have to look for her and find her before you can do much more. When you have found her you can requeen the colony if it is big enough to be able to build up quickly, or you can unite it to another average colony to make use of the bees.

In my opinion the laying-worker colony is a complete loss. It is extremely difficult to requeen, the bees usually killing any queen introduced; the bees are all aged and are of little use to another colony. They will often kill its queen as well. I am afraid my normal method of handling such colonies is to shake the bees on to the ground in front of a big colony and let them work out their own salvation.

If a colony becomes queenless later in the season it may produce laying workers while the beekeeper is waiting for a virgin queen to turn up. As soon as this happens one can be sure that there is no queen and a comb of young brood can be put in for them to make queen cells on. If they do, these can be destroyed, and a good cell put in from one of the other colonies, or from the queen-rearing section, or it is possible to risk introducing a mated laying queen with some chance of success.

Should they refuse to make cells on the introduced comb of brood and no queen can be found, I would place the lot on top of the supers on a big colony and unite them using the paper method (see page 163).

Robbing

This is the nightmare of all beekeepers, because once started it is so very difficult to bring to an end. Two types of robbing occur: that which is usually just termed 'robbing', and 'silent robbing' which is more unusual and more difficult to spot. Silent robbing is when the colony robbing and that being robbed are on completely friendly terms. There is no sign of fighting or unusual behaviour at the entrance; everything is peaceful, but flying will occur when other colonies are all indoors. If it is happening between two different colonies in the same apiary the flight path will be obvious. Often this will go on until the robbed colony is devoid of all stores, when they will starve or possibly all go home to the robber's hive. I always think that this must be the way in which what I call 'Marie Celeste' hives are produced—a hive which is completely empty of bees, stores and brood, but in which every cell is cleaned up and in perfect condition. (With ordinary robbing capping will be present, half torn down, and in the cells from which honey has been removed the coping, or thickening, on the top of the cell walls will be missing, the robbers never stopping to tidy the comb up before they leave.)

Silent robbing is difficult to terminate without taking one colony to another apiary. Changing colonies over by putting the robber in the robbed place and vice versa will often cause sufficient confusion to stop it, but not always. If a second apiary is available I would move the robbed stock away at a time when I could trap as many of the robbers in it as possible. In this way some of the losses can be made up and these bees should help defend the colony in its new apiary.

Ordinary robbing is much easier to spot as the robbers will be flying in a rapid zig-zag fashion in front of the hive, trying to find a way of slipping behind the guards without being challenged. This zig-zag flight alerts the guards and frequent challenges and short flights take place.

Prevention is much better than cure as far as robbing is concerned. The first rule is never to spill honey or syrup about within the reach of bees; never let it drop from supers without cleaning it up; never leave combs with stores in them around where the bees can get at them. This is particularly important as the season advances and every precaution should be taken from the middle of July onwards. Robbing will become an increasing problem as you work colonies late in the season. In August, as you work with open colonies, particularly nuclei, robber bees will follow you around, or rather your smoker, trying to get into the hives—and succeeding. If you carry on working the number of robbers can build up to a level where they are capable of dominating a small colony, and once this happens it is lost.

Reduction of the size of entrance will help to reduce robbing, and as soon as the honey is being removed I would put an entrance block in the big colonies. The nuclei can have their entrances reduced whenever robbers are seen around and if interest begins to build up in the area of the nuclei the entrances can be reduced to one bee way, so that they can do a Horatio act with a better chance. The same thing can be done to any hive that is being robbed, and a good idea is to turn the entrance into a tunnel by using an U-shaped piece of metal about 2 inches long as the only entrance.

If you have left some combs where robbers can get at them, or if they have succeeded in robbing out a nucleus, do not take everything away when you find it happening. Leave a comb with a small amount of honey in it: the robbers will work on this until they have exhausted it and then go home. If you take everything away they will fan out looking for it and may make contact with another small nucleus they can overpower.

Moving the bees to another apiary is again the best answer. In this case I would remove the stock which is doing the robbing if you can identify them, as they will have to reorientate when they get to the new site, which may make them forget the robbing. If you move the robbed stock, the fact that they have already been dominated by another colony and have usually given up defending themselves will make them easy meat to any aggressive stock in the new apiary. As bees are inveterate thieves there is always a number of potential robbers in any apiary.

Disease

Bees suffer from a considerable number of diseases, but we as beekeepers are only interested in a very few. The illness of the individual bee passes unnoticed in the city of many thousands. It is only when epidemic (or more correctly for animals, epizootic) diseases occur that we become interested. When hundreds of bees die we have to do something about it. Equally, we do not wish to harbour disease which may be passed on to our neighbour's bees. It is therefore important that all beekeepers should take steps to inform themselves about the various bee diseases and the methods of dealing with them. The desire of some beekeepers to ignore the matter entirely—even the experienced beekeeper who shuts his eyes to disease in his colonies hoping that it will go away— is deplorable, being both stupid and antisocial.

For convenience, honeybee diseases can be divided into those that affect the adult bee and those that affect the brood. Included in the following are conditions such as starvation, poisoning and chilled brood which, although not infectious diseases may be confused with them by the inexperienced.

Nosema

The causative organism of this disease, *Nosema apis*, is a protozoon, a small single-celled animal like the amoeba, belonging to the Sporozoa. At one period in its life it turns into a spore which is fairly resistant and able to live for several years. The spore is the dispersal form of the animal—the means whereby the disease is spread from one bee to another. The spore is voided in the faeces of an infected bee on to the comb at times when the bees are unable to fly freely. This happens particularly in the autumn, winter and spring. The spores are picked up by the bees cleaning cells ready for the queen to expand her broodnest in the early spring, and some of them are swallowed by the bee and develop in its gut, hatching out and infecting the cells of the walls of the ventriculus. They go through several stages of multiplication and then finally turn again into the spore stage. Heavily infected bees will contain in their gut cells 100,000 spores which are then released with the faeces to carry on the cycle of infection. There are no symptoms which can be easily seen, although there must be some voiding of faeces within the hive to carry on the infection. Nosema is not the cause of dysentry as we know it, but dysentry (see below) is no doubt an efficient method of spreading the disease should it be present.

The effect of nosema on the bee is to shorten its life by about 50 per cent. The effect upon the colony will depend upon the percentage of bees infected. The only practical symptom in the apiary is that the infected colony does not build up in spring and no amount of manipulation will cause it to build up until the disease is reduced in incidence. Colonies with a low percentage of infected bees will not be easily distinguished from colonies which are not affected. Quite heavy infection is needed before the colony is really held in check. However, in betweeen these two kinds of colony there must be many which lose some of their productive capacity. Nosema is not usually a killer in my experience, most colonies recovering from the effects of the disease naturally in about June, when good weather allows all the faeces to be voided in the field and the old infection on the comb has been generally cleaned up as the queen reaches her peak of egg laying. No doubt nosema causes the death of some colonies, but not normally; usually such fatalities occur after a number of consecutive poor summers, and when the bee is being stressed by some additional problem such as dysentry.

The advice I would normally give would be to monitor the presence of nosema spores by a quantitative method if you have the means to do this. Otherwise the service will be done for you on request by local or national advisors. If a rise in incidence is found feed Fumidil 'B' in the

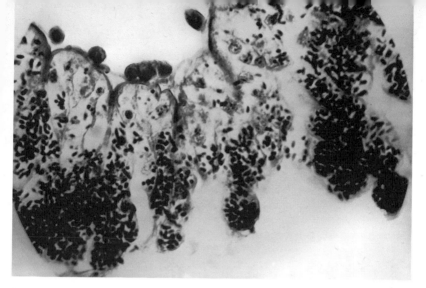

Nosema spores stained black in a section of the ventriculus of the honeybee.

autumn syrup. Fumidil 'B' is an antibiotic used only, as far as I am aware, for the treatment of this disease. It is sold in three-dose bottles and each dose is fed to a colony in 14 lb. of granulated sugar dissolved in seven pints of water. Fumidil 'B' comes in the form of a very fine powder and is extremely, if not impossibly, difficult to stir into syrup. I usually stir it into the dry sugar and then add the warm water (not too hot or it may destroy the Fumidil). The Fumidil syrup is then fed in a Miller feeder or some other rapid feeder so that the colony will store it in a close mass and will therefore live on it for some while. The Fumidil syrup would be roughly the equivalent of 17–18 lb. of stores, and two-thirds of this will see the bees through the first four months after feeding, the remainder being used at the start of brood rearing. This protection reduces the amount of infection laid down on the comb and in my experience nosema is of very little trouble the season after such treatment.

As an extra protection I would suggest that all brood combs empty of brood that are taken from the bees at any time in the year should be sterilized before they are used again in colonies. Sterilization is carried out in the following way. The empty frames of comb are collected into brood chambers, having been cleaned of propolis by scraping the wooden frame; a floor is placed on the ground and a pad of absorbent material into which has soaked $\frac{1}{4}$ pint of acetic acid is laid on it. The brood chamber of frames is placed on top of this, and the entrance is completely closed. If more than one box of combs is to be sterilized a second pad with its $\frac{1}{4}$ pint of acetic acid is placed on the top bars of the frames of the first box. This is repeated at one pad per brood chamber until all the boxes are treated, the top one being covered with a crown board and roof. Some beekeepers cover the pile with polythene sheeting to keep the fumes in. The combs will be sterilized after at least a week in a moderate temperature. The acetic acid you require is the 80 per cent Industrial Grade, which is difficult to obtain in small quantities, and if the beekeeper has to buy the more expensive 'Glacial' Grade, he can dilute this by one part water to every four of acid.

Acetic acid is not a nice substance, and will remove the skin from your fingers in a flash. Rubber gloves should therefore be used when handling it. It will also attack metal and even concrete. It is therefore best to keep the pile of combs being treated outside, away from buildings, and on earth rather concrete. The pile should be examined to ensure that bees cannot get into it as they will rob any honey it contains despite the fumes.

After a week, the combs should be sterile and should be aired for a while to get rid of most of the fumes left in the boxes. The acetic acid does not in any way affect wax or stores, honey or pollen, and all are perfectly safe to give back to the bees. Formalin, which can also be used to sterilize combs, contaminates stores, rendering them poisonous to the bees, so combs treated with this must always be empty and I do not think it really worth trying to use.

Colonies which are affected by nosema in the spring may be treated at this time by first removing the pool of infection which is on the combs not yet used. All combs not containing brood should be removed and sterilized. The colony is then fed Fumidil 'B' to check the disease in the bees themselves. The colony can then be made up with sterilized combs and built up by giving brood from a large colony, as described on page 127.

Amoeba

This protozoon lives in the Malpighian tubules of the bees. It has a resting, distributive stage consisting of a round cyst. Little is known about it and its effect on the bee, but fortunately it is not very common. Fumidil 'B' has no effect upon it but it is killed by the sterilization process mentioned above. From the practical point of view I think we can ignore Amoeba at its present incidence level.

Acarine

Acarapis woodi is a small mite which lives in the main thoracic trachea of the honeybee. The fertilized female migrates into the trachea and begins to lay eggs soon after the bee emerges from its cell. The eggs hatch in about five days and the little larvae, which always remind me of tiny guinea pigs, develop into adult mites about nine days later. The trachea can be stuffed full of mites which feed by piercing the walls of the trachea and sucking the blood of the bee. The trachea are damaged and become brown and brittle, but this seems to have little effect upon

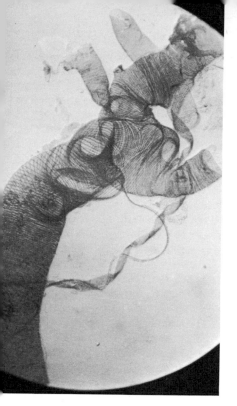

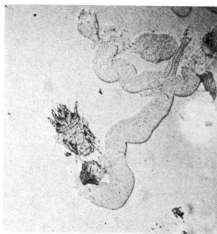

Infestation of the main trachea of the honeybee by Acarapis mites is shown left. The mite's form is clearly seen below.

the bees who can still be working busily: the effect of the mite is probably to reduce the life of the bee somewhat. Some of the mites migrate to other bees as they touch; they do not appear to be able to transfer via the comb or any static object. Having arrived on another bee's thorax they are probably attracted to the wing roots by mechanical vibration and from there they move against the puffs of air coming out of the first thoracic spiracle and enter the trachea.

The effect on the colony will depend upon the percentage of bees carrying the mite, particularly during the winter period, and high infestation may cause the death of the colony. Infestations are high after poor beekeeping summers when bees are confined to the hive and migration of the mites is easy. There are few signs by which the presence of acarine can be detected, but I think that a type of crawling behaviour, where the bees climb grass stems and line up above each other or cluster around the stem, is a sign of bees infested with acarine. In my experience when this type of crawling exists the mite has always been present. Other types of crawling can be caused by many circumstances and are in no way connected with the mite.

The incidence of the infestation varies from area to area in England, and the greatest number of cases is usually found in the West Country and the South with very little in the East, especially the South-east. As mentioned above, incidence also fluctuates with weather conditions and the quality of a year from the bees' point of view—plenty of nectar means a lot of flying and considerable reduction in the number of infested bees.

Never treat a colony for a disease or infestation it is not suffering from. So I would again, as with nosema, try to monitor the disease in my apiary and only treat when required. If colonies are showing no unusual signs of death or reduction in size, or crawling, then all is well. If winter deaths start to increase then microscopic examination will give some idea of the reason. If you have no microscope a sample of about thirty-five dead bees can be sent in a small box to a regional Beekeeping Instructor or the national bee advisors for checking.

If it is established that acarine is present, this can be treated by burning 'Folbex' strip in the hive. These strips of card, approximately 4 × 1 inch, into which is soaked chlorobenzilate, an efficient acaricide, are lit and blown so that they smoulder like a firework touch-paper. The strip is hung in the hive when all the bees are home in the evening and the colony shut in. The bees will immediately fan with a great roar and no doubt the smoke is forced around the inside of the hive to every corner and will be inhaled by the bees into their trachea. The smoke kills the active mites. The dose is usually repeated in a week to ten days so that any mites' eggs present at the first dose will have had time to hatch and be caught by the second. After the hive has been shut in with the smoke for an hour it can be opened to allow the bees to fly if they wish. The treatment does not appear to harm the bees or brood in any way. It is best done when the temperature is above 17°C (62°F) and the bees are showing no inclination to cluster.

The strip must be pinned in the hive in such a way that it is just suspended from a pin and not touching anything else. Where it touches anything the heat will be conducted away and the smouldering edge put out, so that only part of the card will be burnt and only a partial dose given. Usually two doses are sufficient to get rid of the problem.

Paralysis

This disease is caused by a virus which has been given the name of Chronic Bee Paralysis Virus, CBPV. It appears to have many ways of affecting the individual bee and the colony, and its effects were described in the past as several different maladies. The two commonest effects in my experience are the presence of paralysed bees left on the top bars of the frames after the other bees have been smoked down, and the heap of dead and dying bees in front of the hive. In the former case the paralysed bees on the top bar have a flattened appearance, the abdomen may be somewhat bloated, the wings held wider apart than normal and often the whole bee is shivering and shaking. If these bees are prodded they react by trying to raise the abdomen but with little success. Sometimes they have lost some of their hair and look rather greasy. When other bees come into contact with them, they nibble them all over, and sometimes there will be two or three at one time doing this. In my experience, mostly with yellow strains, the disease rarely reaches a worrying proportion, only a score or so of bees being visibly affected at any one time. The worst cases I have ever seen was in a number of dark bee colonies about twenty-five years ago. In this case the dark bees were well-worn and hairless, which made them look small and greasy. Hundreds were on the flight board being nibbled by more normal-looking bees, with more on the ground in a moribund state. There were about a dozen colonies in the apiary and all had this appearance, so much so that at a first glance it appeared to be a massive outbreak of robbing. In the end most of these colonies were wiped out by the disease. This fits the description of maladies in the past which were given the name of 'Little Blacks' and 'Black Robbers'.

The second type of case which seems to be quite common is the one where from 25–100 bees die each day, but leave the hive while in a moribund condition and form a heap of bees below the entrance. The result is sometimes a large heap of dead bees in front of the hive. Beekeepers often mistake this condition for the effect of spray poisoning, but it is easy to distinguish. In spray poisoning the deaths of the flying bees usually occur all at once and it is completely over in half an hour, so the bees in the heap are all of the same degree of freshness, or decomposition, depending on how soon you look at it after the deaths occur. In the paralysis condition, however, a number of bees are dying each day and therefore the heap will be composed of moribund or freshly dead bees on the top and well-decomposed bees underneath. With this type of paralysis the colonies are often very little affected and seem to be able to breed fast enough to keep the population up. From the literature, however, it is clear that many cases have occurred where colonies with paralysis have dwindled badly or died out entirely.

Unfortunately the virus is not controlled by any known drug at the moment, so there is little you can do to help the bees. It has been demonstrated that there is probably a genetic susceptibility to the virus and therefore the usual treatment for bad cases is to requeen with a queen from a different strain. This should also be kept in mind when selecting breeders, eliminating those who are known to produce bees which suffer from paralysis.

Dysentery

This is not an infectious disease as far as we know at the moment, but a malfunction possibly caused by too much water in the gut. This causes extension of the rectum with very fluid faeces which cannot be retained. The cause of the condition is little understood and we can do very little to combat it at present. The condition appears to get worse after several bad honey seasons. In 1968–9 losses in Essex, in southeast England, were very heavy. Many colonies died, with the clusters glued together with faeces. Though the disease is not correlated with nosema in any way it must, however, contribute to the spread of nosema and the reduction of the colonies ability to build up the following season. In the sample I took, about 30 per cent of the colonies had dysentery, and about a third of these died during the winter, of which half had nosema and half were free of this organism. Of all the colonies with dysentery about 65 per cent of them were free of nosema. The sample was too small (about 100) to draw general conclusions.

One of the possible contributing factors towards the existence of dysentery in the winter is crystallized stores of honey. This ties in with the problem in this particular area of south-east England as quite a lot of the honey comes from cruciferous plants: kale, mustard and rape, and crystallized stores are very common. Winter stores of this type can provide the extra water which causes the problem because as the glucose crystallizes out it only takes 10 per cent of the water with it in the crystal; the rest is left with the fructose as a solution between the crystals. This solution can be 4–6 per cent higher in water content than the original honey. The bees will suck this fluid part of the honey from the crystals, often leaving the latter quite dry. The effects of poor seasons could be explained to some extent by the fact that honey is generally of higher water content in the cold wet season. The only advice I would give is always to feed a couple of gallons of sugar syrup per colony in the autumn no matter how much stores the bees already have. If they have no room at all (this is unusual and indicates a poor colony, because the presence of brood should have prevented this amount of storage), remove some frames and put in several empty combs in the middle of the brood chamber. The fact that the ordinary sucrose syrup is stored last means it will be used during the main part of the winter when flying is reduced and dysentry can become a problem.

The problem of dysentery hardly arises in areas with mild winters, which allow bee flight regularly, but a combination of hard winters and an increase of oilseed rape acreage may bring the problem back.

Natural poisoning

This can be caused by plants producing poisonous nectar. This is very rare and I have no personal experience of it. A case did occur in the Isle of Colonsay in Scotland in 1955 when the island was planted with a large number of *Rhododendron thomsonii* which poisoned the bees, killing colonies outright. The West of Scotland College of

Agriculture Study showed that the poison andromedotoxin was involved. Similar problems arise in other parts of the world from other species of plant.

Pesticides

The main poisoning problem comes from the use of agricultural sprays, and considerable damage occurs in most years. The bee can be caught by sprays in three ways: when the crop on which it is working is sprayed, when spray is used on a crop which although not flowering itself, contains a lot of flowering weeds, and when bees are flying over a crop which is being sprayed to reach a forage crop further away. The amount of damage done to the colonies, that is the number of bees killed, will vary with the method of applying the spray. Greatest damage is caused by spraying with fixed-wing aircraft where the blanket of spray will fall from the sky without any warning, and the inability to start spraying directly on the edge of the crop and finish at the other edge may allow the pesticide to fall on areas where bees are working outside the crop area. Helicopters are slightly less deadly as they have more control in this respect, and the down draught from the rotor pushes the pesticide down and at the same time causes enough air turbulence to give the bees a bit of advance warning. They are, however, still deadly if bees are working the crop being sprayed. Finally, the use of tractor-mounted sprayers are least harmful as they do not usually catch bees flying over, and they cause quite a bit of disturbance which will warn some of the insects to fly away.

Time of application is equally important, both in time of day and in relation to the development of the crop. If the rule that no crop should be sprayed when it was in flower was followed, little trouble would occur. But if, through a sudden build up of the pest or, more likely, because of delay in spraying, a crop in bloom must be sprayed, then this must be done when it does least damage: either before 8.00 in the morning or after 8.30 in the evening.

The problem occurs where one or two crops occur, mainly field beans and crucifers such as rape and mustard. The fruit growers, who probably use more sprays than anyone else, cause very little problem; they are so convinced that bees are of use to them for pollination that their system has evolved to a point where bees can be kept near orchards with complete confidence. Unfortunately the loss of bees does not directly affect the farmer, unless he happens also to be a beekeeper, and he therefore does not always take as much trouble as he might to avoid the destruction of bees—although there appears to be a growing appreciation of the beekeepers' problem, which I find very encouraging.

Regarding the two main problem crops mentioned above, field beans was the main crop on which bees were lost for many years. The black aphids turn up in force on the spring-sown beans when they are in flower and the aerial spraying of dimethoate and demeton-methyl against the pest is sure death to any foraging bees working the crop. There is very little justification for causing damage now as granular formulations of pesticides, or selective aphicides such as Pirimcarb, can be used with little danger to the bees. I hope therefore that this problem, which has been a great drain on beekeeping in bean-growing areas, is behind us.

The problem with the cruciferous plants is different. In the past some damage was done to bees where mustard was sprayed for pollen beetle (*Meligithes*) in full bloom. This was unnecessary as the damage is done in the bud and spraying purely for revenge was a waste of money. Oilseed rape is a fairly new crop, certainly in large acreages, and some damage to colonies has already occurred. It is certain that the problem will get worse as the population of the pest, here mainly the seed weevil, builds up year by year, unless some method of dealing with it without killing bees is quickly worked out. Collaboration between farmers, spray contractors, pesticide firms and beekeepers is absolutely necessary, both at national level and between individual farmers and beekeepers at the local level, where the damage occurs. It is to be hoped that the people involved will admit to the honeybee's having some real value and that its preservation will not be dependent upon its not costing anyone anything.

Now let us return to the beekeeping side of the problem: the recognition of spray damage, what to do about it when it does occur, and what can be done to mitigate the problem. As was mentioned under paralysis, beekeepers are often unsure whether deaths at the entrance of the hive are due to poisoning or not. Usually the confusion is between paralysis and poisoning, but even starvation can be confused with these at times. The signs of poisoning by pesticide are usually deaths at the entrance all occurring over a period of thirty minutes to an hour. After this no more deaths occur. The number of dead can vary from must a few to the entire foraging force of the colony: some 15,000 to 30,000 bees, the latter comprising several good shovelfuls of dead bees. If you are in the apiary at the time you will see that many of the returning bees will spin around on the ground until they finally succumb. If they try to get into the hive they will be repelled, and the affected colonies will be extremely upset and nasty-tempered. With paralysis, the bees are dying a few each day for several days, and if they are being nibbled there is not the obvious aggression towards them which is shown towards poisoned bees. Starvation will be shown by bees staggering out of the hive, not with the flattened even-keeled stance of the paralysed bee or the curled-up twitching of the poisoned bee, but bees whose legs do not support them, falling first on one side and then the other.

If you find that your bees have been poisoned, collect a sample of 200–300 bodies, pack them in a cardboard box and post them off to the national authority concerned—in Britain to the National Beekeeper Advisor of ADAS, Ministry of Agriculture. They will analyse the bees for insecticides and it is helpful to provide them with as many details as is possible, if you know them: the crop sprayed, the time of day sprayed, insecticide used, method of application (i.e. aircraft, tractor, etc.) and any other details you think would help. Sending in samples in this way is valuable for two reasons. The results of the analysis could be used to support any claim you make against the person spraying, and your case is added to the statistics of pesticide poisoning which are used to work out ways of preventing such things happening again.

Some areas have spray warning schemes which notify the beekeepers of spraying to occur in forty-eight hours time. Such schemes are very useful as they allow the beekeeper with a few colonies to do something about it, and the large beekeeper to protect such things as queen-rearing apiaries. Not least of all they keep the problem firmly fixed in the minds of those people involved on all sides of the problem who might prefer to forget all about it if possible. Shutting in colonies is very difficult and should certainly not be done in the way used for moving colonies, as this would cause them to heat up and the entire colony to be lost. A method that has been used on a small scale is to throw long cut grass or nettles over the hives, particularly heaping it up loosely over the front. Bees usually manage to tear their way through this fairly quickly, but stay fussing around it rather than flying away to forage. This is an artificially created 'natural catastrophe' to the bees and they deal with it without building up heat and frustration. It is also possible to tent-in a small number of colonies with black polythene, turning day into night but not restricting air flow or the ability of the bees to walk out of the entrance. This sort of thing can be done by the beekeeper with a small number of colonies at the bottom of the garden or in an out-apiary nearby, but is not possible for the larger commercial beekeeper who, with the best will in the world, will not have time to get around his colonies and rig them up before the spraying will be in progress. These beekeepers may easily have several apiaries totalling several hundred colonies at risk at one time.

The answer is more collaboration, better education regarding the use of pesticides, more research into the control of pests in a way that does as little damage to the environment as possible, and good will on all sides.

Starvation

This problem should *never* occur. The beekeeper during routine visits should ensure that colonies have sufficient food for their needs. It should never be assumed that because it is May, June or July colonies can automatically make a living. Many colonies are lost each year because beekeepers think that all must be well at these times, whereas not every colony can manage. In fact in some years little nectar is collected in the early part of the season because of bad weather.

You should be aware of the signs which will occur at the hive when it is starving. Often the first sign will be white pieces of pupae which have been sucked dry before being thrown out. Any time brood is thrown out of the hive the beekeeper should enquire what is going on inside, and one of the causes can be starvation. At other times the first signs of starvation are staggering bees, as mentioned above. They stagger out of the entrance, fall on to the ground and usually stay there fairly still. Looking into the entrance one can see a pile of bees on the floor, either quite still or just feebly moving. If the hive is opened there may still be a few active bees but the majority will be motionless on the comb, many in the cells with just the ponts of their tails sticking out, and some falling down to join those on the floor. The first action is to get a couple of cups of syrup immediately and pour this in the spaces between the combs so that it falls on the bees and then replace the crown board. Within a couple of minutes the bees will begin to revive and in twenty minutes can be flying, throwing out the dead. Once they are in this state a feeder of syrup will give them some stores to play with and the process of building up can commence. But the best thing is not to let this happen by making sure the bees always have enough food.

Brood diseases

We have now to move on to look at those diseases which affect the brood of the honeybee. There are six diseases of this type, of which three are of considerable importance and three are only minor ailments. Of the important ones the first two, American Foul Brood (AFB) and European Foul Brood (EFB), are covered in England and Wales by the Foul Brood Diseases of Bees Order 1967 which gives the Ministry of Agriculture powers to employ inspectors to examine all colonies of honeybee for these two diseases. If they think disease is found they take a sample comb and send it to the laboratory set up for the diagnosis of AFB and EFB. Should the disease be confirmed, a standstill order is issued on the apiary as well as a destruction or treatment order depending on which disease is found. The diseases are not 'notifiable' in the legal sense: that is, if your colonies have the disease and you do not report it to the Ministry you are in no way breaking the law. You must, however, allow the inspectors to examine your bees and you must carry out the directions of the orders should these be issued in regard to your colonies. In most areas the Foul Brood Officers, as the inspectors are usually titled, are great friends of the beekeepers and are often looked to for advice and help in times of difficulty. This helps the Order to run smoothly and means that when disease is suspected by the beekeeper, if he is sensible he immediately gets in touch with the Foul Brood Officer, who is able to deal with the situation efficiently and with the least chance of spreading the infection. In Britain, where the Foul Brood Order has been in existence for some years, it has greatly reduced the amount of foul brood and held it at probably the lowest level of any country in the world.

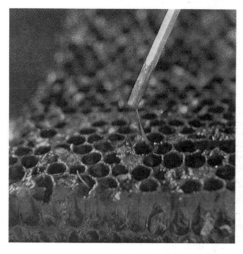

The ropey stage of AFB.

American Foul Brood

The causative organism of this disease is *Bacillus larvae*. As its name denotes, it is a spore-forming bacterium and it is distributed in the spore stage. The spores are fed by nurse bees to larvae, in the gut of which they hatch, becoming rod-shaped bacteria. The rods are the vegetative stage of the organism and the time at which they can multiply in number, although they normally stay more or less dormant in the stomach of the bee larva. Infection usually takes place within the first three days of the larval bee's life; it gets progressively more resistant as it gets older. The bacillus remains dormant until the cell is sealed and the larva is lying along the cell prior to starting its pro-pupal changes. At this time in the larva's life the bacillus breaks out of the stomach into the body cavity where it proliferates, rapidly setting up a septicemia which quickly kills the larva. The larval remains are first yellowish in colour, turning to coffee-brown and then black. During this colour change the consistency of the remains also changes as the whole larva rots down and dries out. At the coffee-coloured stage it has the consistency of a thickish glue. If a matchstick is poked into the cell, stirred and withdrawn, the remains will pull out into a longish slimy strand. This is the well-known 'ropey stage' which is almost a certain diagnostic feature of AFB. The remains continue to darken and dry

'Pepperbox' appearance of AFB. This gets progressively worse as more cells are taken out of circulation because they contain a scale—the queen will very rarely lay where there is a scale.

out and the result is a black scale on the lower horizontal side of the cell finishing about a $\frac{1}{16}$ inch from the edge, and extending slightly up the base of the cell. Due to the consistency of these remains it is impossible for either beekeeper or bee to remove them as they are stuck fast to the wall of the cell. During these changes the bacillus has multiplied enormously and has reverted to the spore stage. Each scale will contain several millions of these spores.

Changes will be apparent to the beekeeper looking at the comb because the cappings above dead larvae become discoloured and greasy looking, and at the same time lose their domed shape and become sunken. Some will be torn down, or partially torn down, by the bees so that perforated cappings is another sign. In the early stages of infection there may only be a few sunken perforated cappings for the beekeeper to see. As the disease progresses, however, more and more cells will contain scales and, as the queen only very rarely lays on a scale, these will remain empty. This means that the brood becomes very patchy, with many empty cells, and this is called the 'pepperbox stage', shown above. When seen, this should always make one suspect AFB. To see the scales, hold the comb with the top bar towards you and then, with the light coming from above and behind, look into the bottom of the cells.

Within the hive infection is spread by the young bees who try to clean up the mess and get contaminated with the spores, which they then pass on to larvae when these are fed. The reduction of available cells in the colony will eventually run down its population until it finally succumbs, probably during the winter, although this may take several seasons.

Infection between colonies is mainly by robbing. When colonies get reduced as described above they are liable to be robbed by a big vigorous colony. Once robbing starts, the flight of the robbers will attract other bees and several colonies may become involved. As the honey in the infected colony will contain spores, and others will be picked up by contact, the infection is rapidly carried home by the robbers and their own colonies become diseased. Fortunately it takes some time for colonies to reach the stage when they can be robbed so that the beekeeper usually sees it first, hence the disease is not very rapidly spread. In most cases it is possible to keep bees for thirty years without ever seeing the disease.

Another way in which infection is spread is by the use of secondhand equipment, and particularly combs. Equipment can be sterilized as detailed below but combs cannot be, and I would never accept combs from another beekeeper unless I was certain of his experience and carefulness. To buy this sort of thing is asking for trouble, unless you are very sure of your man. Infection can be carried over a great number of years by equipment and combs, certainly for as long as thirty-five years, as the spore of AFB is extremely resistant to ageing, to heating and to chemicals.

When a colony is found to have the disease it must be destroyed. The method of doing this is to dig a hole about 3 feet square and deep near to the colony, and place paper and sticks in the bottom ready to start a good fire. Once the colony has stopped flying, it is shut in and a pint of petrol is poured through the feed hole which is then covered. The fumes of the petrol kill the bees within seconds. In the meantime the fire is lit and as soon as it is going well all the frames with the combs in them are put on the fire and the dead bees are carefully brushed in. All the combs and their contents are burnt from both the brood chambers and supers. The hive is carefully scraped clean of propolis and wax and the scrapings added to the fire. The whole hive is then gone over with a blowlamp, singeing the wood to a coffee-brown colour, paying particular attention to getting the heat in corners and crevices. When the combs and bees are completely burnt the hole is filled in with earth, the melancholy job is finished, and it is hoped the disease is completely eradicated from the apiary.

It is always sad to have to destroy colonies, but this is the quickest, cheapest and most reliable way of dealing with this disease. Control by destruction appears to be much more effective than treatment, if one judges by figures from countries where treatment is general. I would therefore always support the continuation of the destruction method for AFB while existing circumstances and methods of medication are unchanged.

AFB is generally distributed over the whole world. There does not appear to be any environmental restriction on its distribution, nor do there appear to be any truly resistant or immune strains of honeybees.

European Foul Brood

This is a very different disease from American Foul Brood. Incidentally the geographical part of the names of these two diseases has no real meaning, being merely the place where they were first written about. European Foul Brood (EFB) is also worldwide but appears to be more local in its distribution, for example in England there was a well-known area in the South-west, in Hants, Dorset and the Wiltshire border country, where it has existed for years, with only sporadic outbreaks in other parts of the country. The position has changed in the last few years and there are small outbreaks in a large number of other regions.

The disease is caused by *Streptococcus pluton*, a very small non-spore-forming bacterium. The bacterium is in the brood food fed to the larvae by the nurse bees, and upon entering the stomach of the larva proliferates and fills the gut, feeding upon the food in the stomach of the larva. It does not penetrate into the body cavity nor poison the larva in any way. If it kills the larva it does so by starving it. Providing the larva is well fed, however, it can take in enough food to feed itself and the streptococci within its gut, and it will then complete its metamorphosis and become an adult. The honeybee larva has a blind stomach as described on page 17, up to the time it is about to become a pro-pupa, when the hindgut breaks through and the contents of the stomach are voided in daubs on the inside of the cell, to be partly covered by the silk of the cocoon. The EFB larva therefore voids thousands of streptococci from its gut, in this way leaving the faeces to infect further occupants of the cell as well as the cleaners. In the early stages of infection few larvae die and those that do are very rapidly thrown out of the hive by the bees, thus removing the infection with the larva. Curiously, therefore the more larvae that die the quicker will the infectious material be reduced, while the more that live to defecate the more infection will build up in the colony. In the early part of the season, when the broodnest is small, the good colony will be able to feed its larvae well and they will in most cases reach maturity. As the broodnest reaches its peak in size towards mid June the amount of infection in the colony will also reach a peak. Nurse bees will be stretched to feed the larvae heavily and some larvae will die and be quickly thrown out. Should there be a sudden reduction in the

EFB : dead larvae showing the typical melted-down appearance, often discoloured or smelly.

forage being brought in a large number of larvae may die at the same time, and the house bees may fail to throw all of them out. This is the time the disease can be seen in the hive, at other times being almost impossible to diagnose.

The larvae generally die before they are ready for the cell to be sealed. That is in the large curled-up stage. The normal healthy larvae are pearly white in colour, neatly curled in the bottom of the cells and exhibiting very little movement. The larvae suffering from EFB turn either slightly yellow or grey in colour and adopt unnatural positions in the cells and show quite a lot of movement—they look as though they are suffering from stomach ache. When dead, the colour becomes more pronounced and the larvae have what has been described as a 'melted down' appearance—rather as though they were made of candle wax which had been subjected to heat. Often at this stage it is possible to see that the tracheal system and the gut may be white, because of its bacterial content. There may be little smell or a very offensive one, these differences being due to the type of secondary bacterial invaders which are helping to rot the larvae down.

Field diagnosis is, therefore, the death of unsealed larvae still in the curled-up position, discoloured and sometimes smelly. Some larvae may make it through to the sealed cell stage when there may be a few discoloured sunken cappings which may be perforated as in AFB. However, the contents of the cell will be quite different in most cases.

Roping does very rarely occur but it is accompanied by a very offensive smell, and the roping will be more granular in texture. Except with the 'ropey' case the dead larvae can be removed whole, or when they have dried down to a scale, which may be positioned anywhere in the cell and is easily removed.

The fact that this disease may be in the hive for some while without visible symptoms, and that when dead larvae do occur they are only present for a very short while before being thrown out, makes the chance of detection by Foul Brood Officers on occasional visits very small. They cannot get around everyone's bees in the three or four weeks when the visible signs are there. The beekeeper should therefore keep a very good lookout for this disease during his routine work and if it is seen or suspected it should be reported immediately.

The destruction of EFB colonies may now be changed to treatment at the request of the beekeeper. In either case, treatment or destruction, the contact colonies (those in the same apiary but not showing the disease) are all treated. Treatment is free and is done by the Foul Brood Officer and consists of feeding a dose of oxytetracycline. Diseased colonies may be treated at any time of year, contacts only in April and May. If the disease is found later than April or May contacts are treated the following year in these two months.

More research on diagnosis and treatment of EFB is needed. How does the disease spread? There has long been a saying, 'AFB by robbing, EFB by drifting'. Anyone with experience of the disease knows that there must be methods of infection other than drifting.

Right *Chalk brood.*

fig. 43 Left *Sac brood: 'Chinese slipper' stage.*

Sac Brood

This has never been a very worrying disease, with usually only a few larvae succumbing to it and no appearance of build up from year to year. The disease is caused by a virus which has been found in honeybees in most areas of the world. The larvae contract this disease, probably from contaminated nurse bees, and they die after their cell is sealed, when they start their pro-pupal moult. The virus appears to affect the process of moulting, preventing the separation of the new and old exoskeleton at the head end and causing large amounts of fluid to occur between the two skins. The result is a tough watery sac, usually greasy at the head end. The larva dies with its head turned up in the entrance to the cell, the bees having removed the capping. This is known as the 'Chinese Slipper' stage and is illustrated above.

There is no known treatment. In very bad cases the best thing to do is to requeen with a queen of a different strain, for as in paralysis there is evidence that some strains have an inherited susceptibility to the disease. Recently there has been more cause for concern, however, as Dr Bailey at Rothamsted has shown that the virus can also affect adult bees, shortening their life and affecting their pollen-collecting capacity. More details are needed before we know how important this effect is in the well-being of colonies.

Chalk Brood

This disease is the result of the larvae eating the spores of the fungus *Ascosphaera apis*. These germinate in the larvae and the mould-like strands, or mycelium, grow until they have completely interwoven the whole body of the larva, which now has the appearance of little fluffy white pieces of cotton wool. However, this appearance is quite deceptive as the little white mummies are quite hard. Some of them will change from white to a bluish-black colour as they become covered with minute black balls—the fruiting bodies containing spores which are eventually set free and dispersed by draughts to cause infection elsewhere.

The disease is not usually a great problem, only the odd one or two affected larvae being present at any one time, though sometimes there is an outbreak where several hundred succumb to the fungus and this is probably in a strain of bee which is very susceptible to the fungus. Requeen with the queen of a different strain is the best advice.

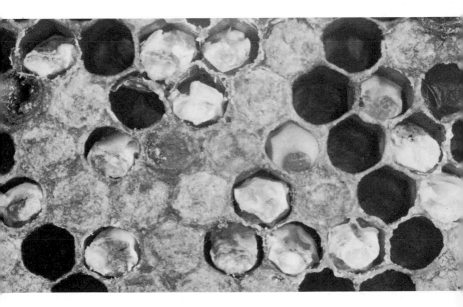

Stone Brood

This is common in Europe and the United States, although only one or two cases have been known in Britain. Like chalk brood, it is caused by a fungus, or rather a number of related fungi belonging to the genus *Aspergillus*. These ramify through the larva or pupa and turn it into a mummy, which instead of looking fluffy and white as in chalk brood appears granular and yellowish. Why the disease should be rare in the British Isles is unknown, because the fungus is present and well distributed, attacking birds and causing Aspergillosis, a disease of the windpipe, and even causing the same in man. Should you get an outbreak of stone brood, do not sniff it or you will yourself be suffering from Aspergillosis. I know of no remedy.

Chilled Brood

Chilled brood can be the accompaniment to starvation, spray poisoning, or mishandling by the beekeeper. Anything which reduces the number of bees below that needed to look after the brood and keep them warm and fed will cause chilling. It is easily recognized because deaths will occur in complete slabs of brood of any age. It cannot be confused with disease, as no disease will infect or kill every individual, nor usually will it be confined to the periphery—the bottom and sides of the brood nest. Prevent this problem occurring by never allowing colonies to reach starvation point, and when doing manipulations, particularly making nuclei, ensure that you do not get the amount of brood and the number of adult bees so inbalanced that there are not enough of the latter to look after the former.

Addled Brood

This is the name given to brood which dies from congenital defects. The queen passes on the factors for these defects to a proportion of her eggs. The proportion will vary from queen to queen and from time to time with the same queen. Mortality can be at any time during the life cycle but we are only likely to notice that which occurs during the brood period. Our notice will be drawn to the condition only when the mortality rate becomes high, as the evidence can be confused with one of the important brood diseases. One type of addled brood can only be differentiated from sac brood by seriological analysis, which few people have the required serum to carry out. However, from the practical point of view this matters little as the treatment is the same in both cases: requeen with a queen of a different strain.

Pests and predators

Fortunately for us, the bee has very few pests and predators for us to worry about in the temperate regions, for it has left behind in its original tropical environments the far less adaptable species that were its original pests and predators. Beekeepers in America suffer predation from skunks and polecats and even more worrying beasts: a lady from the United States visited me once whose main problem seemed to be a large brown bear which lived near by and was also keen on bees. In Europe, problems are on a much smaller scale. I have already dealt with mice and woodpeckers in details of wintering (see page 101). Here I would like to mention other birds and insects.

Swallows and martins will sometimes hawk over an apiary, taking quite a number of bees. Sparrows will on occasion make a dead set at one or two hives and feed their babies entirely on bees. Bluetits will take bees at the hive entrance in summer and may try to get inside during colder weather. None of these does much damage and certainly not to a large apiary, where the total number of bees they take is quite insignificant. However, they are a bit more of a problem in a mating apiary where, I am afraid, they would have to be deterred or got rid of.

The Bee-eater, from its name, a significant predator, but it is fortunately rare, and shrikes also are too rare to worry about.

An insect which I have seen hawking bees is the large yellow and black dragonfly *Cordulegaster boltoni*. I remember one which regularly caught bees in my apiary and took them back on to the bracken to eat; since then I have seen several others of the same species repeating the performance, but never any other species.

The only really important insect pest is the common wasp. These will try to get in the colonies at the end of the season. Starting early in

The victors and the vanquished—workers expelling a wasp from the hive after a fight. The wasp is dead and the worker on the right paralysed by a sting.

August, when the wasp nests begin to break up and the adults go foraging for sugar, the attack builds up in strength until the main population of wasps has died off. During this period any colony which is small enough, or docile enough, to be overcome will be completely killed out. Defence is difficult, particularly of small nuclei. Entrances can be cut down to one beeway or turned into tunnels, which gives the bees an advantage in the fight but may not be enough if the wasps are persistent and numerous. The best defence is attack: find the wasps' nest if possible and kill them out. Do not use cyanide to kill the wasps, you may kill yourself by mistake. I find an aerosol fly killer—one containing a pyrethrum knockdown ingredient—will kill the wasps faster than they can get at me and it is quite safe for humans. If you cannot find the wasps' nest, try placing jars containing a jam and water mixture around the hive. This will divert the wasps' attention and drown them without attracting the bees.

Wax moths

The two wax moths are more damaging to stored comb than to the bees themselves. The species are the Greater Wax Moth (*Galleria mellonella*) and the Lesser Wax Moth (*Achroia grisella*).

The greater wax moth has a wing span of $1-1\frac{1}{2}$ inch and is brownish with varying amounts of ashy-white markings. It has a characteristic way of running from cover to cover with only short flights. The larva, a caterpillar, when fully fed is $1-1\frac{3}{8}$ inch long and pale grey or greyish-yellow, but its head and a chitinous plate on its second segment are reddish-brown in colour. The larva prepares for pupation by finding cover, tucking itself into the space between the frame side bars and the wall of the hive or some such place, and then scooping out a boat-shaped indentation in the wood. It spins a cocoon in the trough so formed and pupates. The moths are somewhat gregarious, and have the instinct to pupate side by side, touching each other, and as each has its trough scooped out in the wood the amount of damage they do to equipment as well as comb is quite considerable. The main damage is to stored comb, and particularly any which has been bred in, the bees faecal matter and old cocoons supplying some of the necessary parts of the caterpillars' diet. Brood comb is therefore particularly at risk. I have seen two brood chambers full of old brood comb completely turned into silk and dust in fourteen days, in a position where beekeeping personnel were passing every day and not a moth was seen. Had it been left any longer there would of course have been moths everywhere, as the new generation arrived. Wax moth larvae do not appear to like honey and as super combs are often stored wet from the extractor, they are in my experience reasonably safe from damage, providing they are not too old and have not been bred in. Treatment of stored comb is dealt with below.

Above *The silken tunnels of the greater wax moth running through the cells of the comb. Had this been left undiscovered the larvae would have gradually eaten up the wax, leaving only the wood and the wire.*

Right *Two greater wax moths. These are probably males: the female usually has lighter underwings and a different wing profile.*

The other bad habit of the greater wax moth is to get into the combs with brood in them in the hive. The moth is so skulking in its habits that it manages to escape the bees and to lay eggs amongst the brood. These hatch out and will tunnel along the cappings of the brood leaving, when they have grown a bit, a white line of silk, about 2 mm wide, covering the tunnel. These lines may be chased along with the corner of the hive tool until the larva is discovered and killed. It has been shown that the faeces of the wax moth larvae can, when left in the cell, affect the bees' pupation, producing abnormal adults with shortened bodies. The number of wax moths which will be found in a colony will depend upon the housekeeping capabilities of the strain or race of honeybee. In my experience the Italian strains do not tolerate company of this sort and only the odd larva will be found in their colonies. Some of the darker bees of the Northern European race will put up with considerable invasions of wax moths, and I have seen some in which there were almost as many moths as bees. If wax moths living in the colonies with the bees starts to become general I would suggest changing the strain of bee for a better housekeeping strain.

The lesser wax moth has a wing span of $\frac{5}{8}$–$\frac{7}{8}$ inch, and is silvery in colour with no pattern markings. It runs about very quickly and flaps its wings very rapidly while hanging on to the comb. The larvae are somewhat smaller than those of *Galleria* but otherwise look identical in colour. They are more gregarious in feeding and produce a large web of silk in which they live, extending the web as comb is devoured.

This species will eat the comb quite readily but does not damage the woodwork when pupating. Neither do the moths live in the combs amongst the brood of the bees.

Treatment of stored comb

Probably the best way to store dry comb to prevent wax moth damage is to place a sheet of newspaper in the piles of supers at every fourth super. A teaspoonful of PDB (paradichloro-benzene) is placed on each paper sheet inside the stack. PDB will control the larvae of the wax moth so that any combs, dry supers or bred-in brood combs, can be stored in this way. It is not advisable to store supers in this way in a lived-in room of the house, as the vapour of PDB will give one a nasty headache if too much time is spent in contact with it.

Combs should never be sprayed with an aerosol moth killer, nor should insecticide dispensers of any sort be allowed near comb. Beeswax will absorb many insecticides very rapidly and will be fatal to the colony the following season. This can also happen to combs stored in a building during, and for some time after, it is being treated to destroy wood-boring beetles.

Comb can be stored wet with honey, particularly where a large number of supers are involved and where buildings are kept free from

Braula coeca, the bee louse, riding like a tiny red jockey on the back of a bee.

any build up of moths. This method does not find favour with everyone because the residue of honey will either crystallize, thus providing a starter to increase the chance of crystallization of the next year's crop, or it will deliquesce, make a mess, and build up yeasts in the storage areas.

Braula coeca

The 'bee louse', *Braula coeca*, is in reality a true fly, a member of the order Diptera. This species has during evolution lost the normal pair of wings. It is an interesting little animal which lives in many colonies of bees, the adult riding around on the backs of the workers and often more particularly on the queen. They have the appearance of tiny red mites each about the size of a pin's head. Often a dozen or more will collect on the queen. As far as we know they do not harm the bee at all, merely riding around on them and joining in when two bees are transferring food. They do not suck the bee's blood or damage it in any way.

The eggs of *Braula* are laid in the capping of the honey and the little larvae tunnel along on the underside of the capping, cutting a channel in the inside face of the capping and covering this with silk. The effect is a tunnel that runs along in the capping rather like a leaf miner in various plants. Presumably they live on the wax, pollen and honey which is available to them, the pollen being mixed in the capping.

They only really worry beekeepers who are showing sections or cut comb. To these people they are a nuisance as they weaken the capping over the honey and let the latter ooze out to deliquesce and generally make a mess over the surface of the comb—what the beekeeper calls 'weeping sections'.

I never worry about *Braula* unless the queen gets covered with them, and then I remove them by taking the queen in my cupped hands and blowing cigarette smoke over her. The *Braula* immediately fall off the queen on to the palm of the hand. The queen can then be returned to the colony free of 'lice' and the *Braula* destroyed. I know of no reliable way of eradicating them from the whole of a colony.

Mouldy comb, mouldy pollen and pollen mites

Small colonies which cannot control the internal environment of their hive, or colonies which are in extremely wet winter apiaries, will often have mouldy combs on both flanks of the hive in spring. These areas of mould are not readily used by the bees again and combs of this sort should be removed and melted down to salvage the wax. Prevention is better than cure and therefore dry sites, strong colonies and plenty of stores should be ensured to prevent mouldy combs appearing.

In the winter or early spring the cappings of stores will have a 'bloom' on them which worries some beekeepers, who may think this is what is meant by mouldy comb. This bloom does no harm and is nearly always present in winter combs.

Mouldy pollen is caused by the fungus *Ascosphaera alvei*, a close relation of the cause of chalk brood. The mycelium of the fungus ramifies through the cell of pollen, turning it into a hard block. It is then useless to the bees and cannot be removed by them without tearing down the cells. A small patch can be cleaned out by the beekeeper, by scraping to the foundation, but large patches will make the comb useless and therefore it is best to remove the comb and salvage the wax. If colonies are well supplied with stores any pollen present in the comb will be covered with honey and capped over for winter and the problem of mouldy pollen will not arise.

Pollen mites may worry the beekeeper who has never seen them before. The usual sign is that in stored combs pollen will turn into a sawdust-like substance, spilling out from the cell. If it is examined carefully it can be seen to move, and with a hand lens little hairy mites will be seen moving about on it. Do not worry: these do no harm and if given time will turn the dry stored pollen into dust, which can be shaken out by the beekeeper, or cleaned out by the bees, thus preventing its being attacked by *Ascosphaera* and turning into the much more tiresome and irremovable stone of pollen mould.

10 Flowers for food

No one can think of bees very long without thinking also of flowers. The coloured flowers and primitive bees probably evolved together, the bee getting its food from the flower and attracted by its colour, the flower advertising its wares with colour to ensure its pollination.

The normal sugar found in the sap of plants is sucrose. Sucrose is a molecule made up by the chemical combination of two simple sugars, glucose and fructose, or using their older names, dextrose and levulose. These sugars are termed monosaccharides, and a sugar such as sucrose which is formed by the combination of two monosaccharides is termed a disaccharide. These sugars are secreted by the nectaries in the flower in the form of nectar—a sugar and water solution. These nectaries are not just holes in the plant which allow sap to escape but active organs which select from the sap those substances which are to be secreted as nectar, and in some species the sucrose is partially or even wholly broken down into its component monosaccharides before it is secreted as a nectar. Nectar can vary, therefore, in containing either pure sucrose, a mixture of sucrose, glucose and fructose, or just the two monosaccharides, glucose and fructose. As examples, *Ribes* (black and red currant) are in the first group, most of the clovers in the second, and the brassicas (rape, mustard, kale) in the last one. In the last flowers glucose is in excess of fructose, and this is the reason why the honey from these plants granulates very quickly, often in the combs.

The nectaries occur in many places in flowers, but usually near the base of the stamens. Depending upon the construction of the flower they can be exposed, as in apple, or concealed as in clover, which is a much more complex shape. If the nectary is concealed, the nectar is not as subject to alteration by environmental effects, such as wind and rain, and it needs a long-tongued insect to find its way into the flower, which means some sort of bee.

Michaelmas daisies. Bees seek them out for nectar.

As we have seen, different species of flower will secrete different types of nectar, and they also vary in the average concentration of nectar they produce. These two factors will affect their attractiveness to the bee. This is something which is important in pollination work, as 'crop competition' will often have adverse results. For instance, apple, producing nectar at 25 per cent sugar cannot keep the bees interested if there is a kale field down the road secreting nectar at 50 per cent sugar. Also in this case the type of sugar in the nectar will also make the kale more attractive to the honeybee. The percentage of sugar in an average nectar for a given species can vary from about 5 per cent for the primrose to 70 per cent for white horsechestnut.

Variation in sugar concentration in, and quantity of, nectar secreted will vary in the same species between plants in different habitats, and in the same plant under varying conditions. Plants do best in the habitat for which they have evolved: heather requires an acid soil and a reasonably high rainfall; white clover does best on the chalk, limestones or the alkaline clays. There is also considerable variation in the yield of individual species—some plants suffer more than others from the effects of drought. This is something which happens in some years to white clovers on thin chalk soils when the same species on 'clay with flints' or 'chalky boulder clays' is still producing extremely well. There are short-term fluctuations: cold weather will reduce the rate of nectar secretion or even stop it altogether; warm weather increases the rate of secretion and very hot weather will stop it again as the plant wilts. Rain may wash the nectar from an open nectary as in apple, whilst sun and light wind may dry the water from the nectar causing the sugar concentration to rise. Thus an apple tree can produce nectar at 25 per cent on one side and 52 per cent on the other, sunny, side.

All these factors affect the amount of honey a colony will be able to collect and store. They are of interest to the practical beekeeper when moving bees to crops or picking out-apiaries away from home. The density of forage in a district, the suitability of the plants to the area, and the probability of good weather conditions will decide how many colonies can be kept in a district without overstocking it. The amount of nectar in a district is a finite quantity: it can all be collected by X colonies each giving a good crop, or by 2X colonies each giving half the crop, but with double the work and double the capital outlay. Care should also be taken not to reduce the crop of a beekeeper already in the district.

The successful beekeeper has to be something of a botanist as well. He should get to know the flowers of the countryside and the crops grown, as well as the methods of pest and weed control employed. This not only makes beekeeping a great deal more interesting, but will help you to get more honey and save you from some of the heartaches of having colonies killed by spray poisoning.

There follows a list of the main honey forage plants of the British Isles, many of which will be found in other countries as well. These plants are all ones which, if they are present in quantity, will produce a honey crop on their own. The figures for colony density given for some of the field crops are the number of colonies per acre which should gather 100 lb. (45 kilos) of honey or more, providing the field crop is well grown and the weather is favourable. A second list starting on page 224 gives further very useful forage plants, some of which help spring build-up. Some of these will provide a honey crop in their own right, and some will provide very highly flavoured honeys which can be blended with some of the less flavoured honeys which come in much greater quantity. Pollen-load colours are given because the beekeeper should learn to recognize these: they will give him information about the forage being worked by his colonies.

Major forage plants

Acer *see* Sycamore

Aesculus *see* Horsechestnut

Alfalfa *see* Lucerne

Bean *see* Field Bean

Blackberry *Rubus fruticosus* agg. (Rosaceae). This ubiquitous plant is made up of a large collection of different species, which tolerate different soils and conditions, and have very long flowering periods stretching from June to August. Well worked by bees even at fairly low temperatures, supplying both nectar and pollen in quantity. Honey of good flavour, medium amber, tending to granulate with a coarse-grained texture. Pollen load pale brownish grey.

Blackcurrant *see* Fruit, soft

Brassica *see* Rape

Broad Bean *see* Field Bean

Calluna vulgaris *see* Ling

Chamaenerion angustifolium *see* Willowherb

Charlock *Sinapis arvensis* (Cruciferae). This weed of arable land was once very widespread but is now held severely in check by the hormone weedicides. In mass it is slightly paler yellow than mustard or rape. It is extremely attractive to the honeybee and highly productive of both nectar and pollen. The honey granulates very quickly, often in the comb, is white or extremely light straw coloured, with normally a fine granulation. Pollen loads are clear yellow.

Cherry *see* Fruit, top

Clover, red *Trifolium pratense* (Papilionaceae). There are a number of varieties of red clover grown, but none of these is in my experience of any use to the honeybee except those grown as 'double cut' clover—first grown for a crop of silage and then, when the plant has recovered, for a seed crop taken at the end of the season. The first growth has far too long a flower for the honeybee to be able to reach the nectar, but the second-cut flower is shorter and once the nectar has risen in the neck of the flower it can be reached by the bee, and once she can reach it she seems to be able to take the lot out. The clover does best on a chalk soil and requires several days over 22°C (71°F) before it begins to yield. Once the flow has started it is the biggest and most rapid that I have ever experienced. The honey is water white, granulates within a couple of months of extraction and with a rather mealy texture. No flavour except sweetness. Pollen load brown. Colony density 2½ per acre. Flowers from mid July to end August.

Clover, white *Trifolium repens* (Papilionaceae). This is the honeybee plant *par excellence* and up to a few years ago produced the main proportion of honey in Britain. This was particularly so with the small wild white variety which grew in the old permanent pastures. The new varieties which grow more lushly, being selected to produce leaf, are not as good nectar producers as the wild variety. The plant likes alkaline soil conditions and prefers the chalk; it rarely seems to produce much on the alluvium in valley bottoms nor does it grow well in wet conditions. It suffers during drought conditions, and in dry years does best on the alkaline clays. It yields best when temperatures are 20°C (68°F) and above. The quantity of white clover has fallen with the loss of the old permanent pastures and the increased use of nitrogen on grass. The honey is pale straw colour and granulates within a couple of months of extraction. The texture of the granulated honey is fine but very hard and often has a glassy appearance. The flavour is exceptionally good. Pollen loads brown. Colony density 1 per acre. Flowers mid June to end July.

Crataegus *see* Fruit, top

Dandelion *Taraxacum officinale* (Compositae). This well-known yellow daisy flowers in verges, rough bits of undisturbed ground and permanent pastures, often in some profusion. It is an excellent honeybee flower, producing both nectar and pollen in profusion, and at fairly low temperatures. The nectar flow is usually recognizable because the capping wax is a light lemon yellow. The honey is golden, of good flavour but granulates somewhat coarsely. Pollen load bright orange. Flowering mainly mid April to mid May.

Erica species *see* Heather

White clover. The wild strains produce the greatest quantities of honey.

Field Bean *Vicia faba* (Papilionaceae). The species includes the broad bean as well as the various varieties of field bean. The latter is an important break crop in some regions and is of two main types: that sown in the autumn, flowering the following May, and the spring-sown varieties flowering in June and the beginning of July. The autumn-sown bean yields well and more regularly than the spring-sown varieties, and is it not attacked by the black aphid as it has finished before the aphid population is any size. Thus there is not normally a spray problem to the bee. The spring-sown varieties are

The field bean produces a very richly flavoured honey in moderate quantity.

much more likely to be attacked by the aphid and have constituted the main risk of spray-poisoning to the honeybee over the last twenty years or so. The position is improving where farmers make use of the less destructive methods and sprays. These varieties do not yield nectar as regularly as the autumn sown bean and often fail to yield at all. The honey is medium amber, of rich flavour and medium-grained granulation. Pollen load slate grey or yellow brown to brown. Colony density $\frac{1}{2}$ per acre.

Fireweed *see* Willowherb

Fruit, soft This includes black and red currant and gooseberry, *Ribes nigrum*, *R. rubrum*, and *R. uva-crispa* (Grossulariaceae). These are grown in various places on a commercial scale and provide both nectar and pollen for the honeybee. They flower early in the year—the end of March and first half of April, the gooseberry preceeding the others—and therefore are useful in the build-up period. Honey pale and of mild flavour. Pollen loads greenish grey.

Fruit, top All the tree fruits—apple, pear, plum, cherry and hawthorn or may—are lumped together under this heading. Thus *Malus, Pyrus, Prunus* and *Crataegus* (Rosaceae) are included. None of these plants is a good honey plant, but they provide some nectar and plenty of pollen and are therefore useful to the beekeeper. The flowering periods cover the end of March, April and May. The flowers are not particularly attractive, producing nectar of low sugar concentration. The honey is pale straw to medium amber and of delicate flavour. Pollen loads greenish yellow to pale yellow.

Gooseberry *see* Fruit, soft

Hawthorn *see* Fruit, top

Heather *Erica tetralix* and *E. cinerea* (Ericaceae). These flowers are found on acid soils and mainly in the heaths and on moorland. Their yields are fairly irregular but can be quite large. I understand 1975 was probably the best year for bell heather in Scotland this century. The honey is 'port wine' in colour and is a fluid being extracted in the normal way. It is strong in aroma and taste, being quite bitter. Flowering period end July and August. The honey from these plants is not usually classed as 'heather honey'. This is produced from ling.

Horsechestnut *Aesculus hippocastanum* (Hippocastanaceae). This large ornamental tree is commonly grown in Europe and is well known by its large palmate leaves and conspicuous upright white inflorescences, commonly called 'candles'. The white horsechestnut is very attractive to the honeybee, providing both nectar and pollen in good quantity. The honey is light and granulates smoothly. Pollen loads are red. Flowering period April. The red horsechestnut is not as common and not so attractive to the bees. It has even been known to poison bees at times, although this is usually bumble bees.

Lime *Tilia* species (Tiliaceae). There are a number of species of lime, two of which have been known to poison bees in some seasons: *Tilia petiolaris* and *T. orbicularis*. The two common limes are *T. vulgaris* and *T. platyphyllos* and fortunately they do not appear to affect the bee adversely. Lime has always been a great mystery to me; I have not seen any quantity of lime honey since the early thirties, when a friend of mine used to get quite a lot from a few trees near his apiary. Since then I have had apiaries by the side of large avenues of limes and failed to detect any sign of honey from them. In some regions there are good yields and a particularly good area in England is near Birmingham. Reading old literature suggests that the species provided more honey in the past than it does today. The honey is pale straw with a slight greenish tinge. The flavour is delicately minty. Flowering occurs about the end of June. Pollen load colour pale greenish yellow.

Ling *Calluna vulgaris* (Ericaceae). This is the source of a glorious and unusual heather honey. It is a plant of acid lands and of the heaths and moors. There is a myth in beekeeping literature that this plant only produces honey if it is growing above 1,000 feet above sea level. This is nonsense as, for instance, the New Forest in England produces a good flow most years and a good, pure sample of ling honey on its own. The only excuse for the myth is that at upper heights there is less chance of other honeys being collected as well as the ling. The flow is fairly irregular, good years coming one in every three to five years, but usually there is some crop. The honey is quite unlike any other being a thixotropic gel, i.e. it can be turned into a fluid by stirring but returns to a gel again when left alone, and it has an aroma and bitter flavour all of its own: it is a truly glorious honey. Pollen load colour is grey to greyish-white. Flowering period mid August to early September.

Lucerne Alfalfa *Medicago sativa* (Papilionaceae). Grown as a fodder crop in many areas and cut before flowering. However, when a field is allowed to flower a crop of nectar will be taken by the bees. The flower likes it warm, around 21°C (70°F), for a good flow of nectar. The honey is light and very mild in flavour, and fine in granulation texture.

Malus *see* Fruit, top

May *see* Fruit, top

Medicago *see* Lucerne

Melilotus *see* Melilot

Melilot (Sweet Clover) *Melilotus alba* and *M. officinalis* (Papilionaceae). This plant is still grown in some areas as a seed crop, providing a reasonable crop of honey. The honey is light, with a good flavour, and when granulated it is finely textured.

Mustard, white *Sinapsis alba* (Cruciferae). This is the crucifer with the brightest yellow flowers. It is grown for seed to supply the ground mustard used at the table. It is also grown quite extensively for ploughing in for green manuring. Like all the crucifers it is extremely attractive to the bees, producing copious nectar at quite low temperatures, 18°C (65°F), and large amounts of pollen. It has been a crop which has produced the second highest problem from spray poisoning, when growers were attempting to control pollen beetle. The honey is white, with a poor flavour and aroma. It granulates very quickly in the comb when taken from the bees, and granulation is very

Ling, source of heather honey.

fine. Pollen loads bright yellow. Colony density 1 per acre. Flowering period May–June for seed, August–September for green manuring.

Oak *Quercus robur* and *Q. peduncularis* (Fagaceae). These trees are listed not because they supply much nectar or pollen, although the honeybee does work the catkins of oaks in some years, but rather because they are a very wide source of honeydew produced by the aphids living on them and this is harvested by the bees just like nectar. Honeydew can vary from golden to almost black, and usually granulates with difficulty, forming large sheets of coarse crystals against the inside of the container. If induced to granulate by seeding the result is usually of a brown shade. The flavour is much prized and is often rather like the taste of figs. Produced mid and late season.

Oilseed Rape *see* Rape

Pear *see* Fruit, top

Prunus *see* Fruit, top

Pyrus *see* Fruit, top

Quercus *see* Oak

Rape Oilseed Rape *Brassica* species (Cruciferae). This is a crop which is fairly new to Britain and is increasing in acreage rapidly. The yellow flowers are slightly paler than mustard but are possibly more attractive to the honeybee, good crops of honey being obtainable wherever the flower is grown. Unfortunately it is potentially a catastrophe as far as the honeybee is concerned if the crop is sprayed against the seed weevil. The flower is so attractive that bees will come from at least two miles all around to work it and the colonies reaching sprayed crops will be ruined for the rest of the season, whilst the wild bees will be totally decimated. The honey is white with poor flavour and aroma, and granulates the fastest of all the crucifer honeys, so extraction should begin as soon as the flower goes. Pollen production is good and the loads bright yellow. Production May, June and part of July for the different varieties. Colony density 2–3 per acre.

Raspberry *Rubus idaeus* (Rosaceae). A very attractive crop to the honeybee because of high sugar concentration of nectar. There is considerable acreage in Eastern Scotland around Dundee. Honey is light with fine delicate flavour and aroma, granulating very quickly and finely. Pollen load greyish-white. Colony density 1 per acre.

Red Currant *see* Fruit, soft

Ribes *see* Fruit, soft

Rosebay *see* Willowherb

Rubus fruticosus *see* Blackberry

Rubus idaeus *see* Raspberry

Salix *see* Willow

Sinapis alba *see* Mustard

Sinapis arvensis *see* Charlock

Sweet Clover *see* Melilot

Sycamore *Acer pseudoplatanus* (Aceraceae). A very good nectar tree, being well worked by the honeybee who finds copious nectar secretion and plenty of pollen. Honey is light amber with a greenish tinge, good flavour and granulation. Pollen load greenish grey. Flowers in May.

Taraxacum *see* Dandelion

Tilia *see* Lime

Trifolium pratense *see* Clover, red

Trifolium repens *see* Clover, white

Vicia faba *see* Field bean

Willow *Salix* species (Salicaceae). This genus has a very large number of species all of which produce nectar and pollen. The plants are dioecious, having male and female catkins on different trees. Male trees therefore produce both pollen and nectar while the female tree only nectar. This is one of the few shrubs which is a real must in the beekeeper's garden, the golden variety, with large catkins, of *Salix caprea* being the best, see page 34, providing a wonderful supply of

A coming and going of bees on rosebay willowherb, profuse source of nectar and pollen.

early pollen to the bees. The species flowers from March to May and is therefore too early to yield a honey crop, but they are of great value for spring build up, often being the plant that really gets the bee year going. Pollen is clear yellow to slightly orange-yellow and in prodigious quantities.

Willowherb Fireweed, Rosebay *Chamaenerion angustifolium* (Onagraceae). An excellent honey plant which grows in profusion in various places, particularly where there has been fire or where trees have been felled. The tall racemes of rose-red flowers have a long flowering period and are usually well crowded together. The output of nectar is very high. The honey is white or very pale straw in colour, the flavour is exceptionally good—will always take prizes on the show bench—and the granulation quick and fine textured. Pollen load is blue or greenish-blue. Colony density 2–3 per acre.

Forage plants of secondary or local importance

Acer campestris *see* Field maple

Acer platanoides *see* Norway maple

Acacia The true acacias, such as mimosa, are found in the warmer climates such as Africa and Australia.

Acacia *see Robinia*

Balsam *Impatiens glandulifera* (Balsaminaceae). This and other species occur in Europe and Africa and are useful forage plants for nectar. They are naturalized over most of the British river courses. There seem to be some individual differences as some plants are worked heavily by the bees and others ignored.

Birdsfoot-trefoil *Lotus corniculatus* (Papilionaceae). This red and yellow flowered plant with pea-type flowers is a very good bee plant. It occurs generally in Europe over the country in waste places and on road verges. It has also been grown as a fodder plant.

Bluebell *Endymion non-scriptus* (Liliaceae). Large bluebell woods still occur in some areas. The honeybee works this flower with varying eagerness but in some years with great vigour. It provides a yellow pollen.

Broom *Sarothamnus scoparius* (Papilionaceae). This plant is well distributed, growing on acid soils. It produces large quantities of orange-brown pollen fairly early in the season. There is argument as to its production of nectar; I think it is little worked for nectar, although some of the garden varieties are.

Bryonia *see* White bryony

Buckwheat *Fagopyrum esculentum* (Polygonaceae). This plant is well known in Europe. The honeybee works it readily and the honey, produced in considerable quantity, is dark and strong-flavoured. Pollen yellow. Colony density 1 per acre.

Castanea sativa *see* Sweet chestnut

Centaurea nigra *see* Knapweed

Citrus Citrus fruits provide some excellent honeys in various parts of the world, the best known ones being orange honey from Spain and California.

Clematis vitalba *see* Old man's beard

Comfrey *Symphytum offininale* (Boragineaceae). Liking damp patches, comfrey is quite common in ditches. There is a little grown as fodder in Britain but it is much more widespread in Europe. The plant provides both honey and pollen.

Corylus *see* Hazel

Echium *see* Viper's bugloss

Elm *Ulmus* species (Ulmaceae). This tree flowers early in March and provides an early pollen flow. In some areas beekeepers regard it as a major factor in helping to start the colonies building up in spring, but with the ravages of Dutch elm disease in the last few years its importance must be on the wane. Pollen loads grey to white.

Endymion non-scriptus *see* Bluebell

Eucalyptus Trees of this genus, of which there are a large number of species, are the main honey producers in Australia. The honey varies in quality from very good to very poor, but always, however, in quantity. In the temperate latitudes of Northern Europe one or two of the hardier varieties are grown and can get to good size, and in the warm summer these will produce a large amount of nectar and pollen.

Fagopyrum *see* Buckwheat

False Acacia *see Robinia*

Field Maple *Acer campestris* (Aceraceae). This can be a small nicely-shaped tree but is more usually a hedge plant. Bees seem to work it mainly in the morning after a heavy dew, and it appears to produce a good quantity of nectar.

Gorse *Ulex europaeus, U. gallii, U. minor* (Papilionaceae). Extremely spiny large shrub with brilliant golden pea-type flowers. The first of the three species named flowers from March to June and the other two July to September. They probably produce a little nectar but are mainly pollen plants, the pollen loads being brown in colour. A very useful species to aid in spring build-up.

Hardhead *see* Knapweed

Hazel *Corylus avellana*. This tree produces its long 'lambs tail' catkins in February to April, there being a considerable difference in the flowering time of various individual plants. It is purely a pollen plant as far as the honeybee is concerned and is worked avidly if the weather allows. The pollen has a fairly low nutritive value but is usually the first fresh supply available. Pollen load pale greenish-yellow.

Hedera helix *see* Ivy

Holly *Ilex aquifolium* (Aquifoliaceae). This well known bush with the prickly leaves and red berries is a very good bee plant. It is on the whole dioecious, and therefore some plants will supply only nectar. Nectar is produced in considerable quantities and eagerly worked by the bee. Flowering period is May to June. Pollen load greenish-yellow.

Ilex *see* Holly

Hazel catkins (above) are an early source of pollen if the weather is warm enough for the bees to fly, and crocus (below) is another excellent source of early pollen. Low on the ground, and therefore sheltered, it can be worked at low temperatures. Pollen is carried on the hind legs of the worker honeybee, held in by the hairs which form the pollen basket, or corbicula. A partial load is shown left.

A good autumn source of both pollen and nectar, with a long flowering period is the blackberry. In poor soil areas it is often the main forage plant.

Impatiens *see* Balsam

Ivy *Hedera helix* (Araliaceae). Ivy is the producer of the last nectar and pollen flow of the year, being in bloom from the end of September to October. When the weather is favourable the honeybee works ivy with considerable avidity. In some parts of south-west Wales and Eire quite a lot of colonies' winter stores are expected to come from this plant, the climate making it workable in most years. Problems have been caused by this as the white honey very rapidly granulates and has been known to dry out so that the bee was unable to use it during the winter. There have also been cases of bees being found with the honey granulated in their honey stomachs, but whether this is cause or effect of death is not known for certain. Pollen loads are yellow-orange.

Kale *Brassica* species (Cruciferae). The kales are widely grown for seed in the East of England, mainly in Essex, and like all the crucifers are extremely attractive to the bee. The honey very rapidly granulates and is usually white in colour and poorish in flavour. Masses of pollen is produced, the loads being yellow. The plant competes for the bees with the fruit orchards, and usually wins, being out in April-May.

Knapweed Hardhead *Centaurea nigra* (Compositae). This is a flower of road verges, rough areas, heathland, etc. It flowers in July and August and looks rather like a soft non-prickly thistle. The bees work it readily for both nectar and pollen. The honey is dark and very strong in flavour—like cough mixture—and is excellent for blending. Pollen loads greyish-white.

Lavender *Lavandula* species (Labiatae). Grown for its well known scent in most gardens, these plants provide excellent forage for the bee. There are considerable acreages grown in Europe and migration of bee colonies to the lavender fields is an annual event. The honey is medium to dark amber in colour and strongly flavoured.

Lavandula *see* Lavender

Limonium *see* Sea lavender

Lotus *see* Birdsfoot-trefoil

Ligustrum *see* Privet

Maple *see* under Field or Norway Maple

Marjoram *Origanum vulgare* (Labiatae). A chalk-land flower of considerable value. The honey is light with a greenish tinge, the flavour very minty and quite delicious. Pollen loads are brownish-grey.

Norway Maple *Acer platanoides* (Aceraceae) A medium-sized tree rather like the sycamore but instead of the pendulous panicles of the latter those of Norway Maple are erect and the flowers are out before the leaves in April-May. The bees work the tree avidly if weather permits.

Old Man's Beard *Clematis vitalba* (Ranunculaceae). Also known as traveller's joy, this plant has become more common with the advent of mechanical hedge clipping. It does best on an alkaline soil and is particularly evident upon the chalk. It is worked quite heavily by the bees for both nectar and pollen, the latter being produced in abundance. Pollen loads white, flowers July and August.

Onobrychis *see* Sainfoin

Origanum *see* Marjoram

Papaver *see* Poppy

Phaseolus *see* Runner bean

Poppy Field Poppy *Papaver rhoeas* (Papaveraceae). The ordinary red field poppy is principally a plant of the arable areas and favours the chalk lands. Only producing pollen, with no nectar, the bees seek it out as it is for some reason extremely attractive to them. Pollen is produced in enormous quantity and when loaded, or stored, appears quite black. It is, however, very dark blue on closer examination.

Privet *Ligustrum vulgare, L. ovalifolium* (Oleaceae). This common hedging plant also grows wild in some areas where it provides berries for the birds. The bees work the white erect panicles of flowers for both nectar and pollen. The honey has the reputation of being objectionable in smell but I cannot say I have ever found much wrong with it. The honey is dark, with a very strong flavour. Pollen loads pale yellow.

Ragwort *Senecio squalidus, S. jacobaea* (Compositae). Both these plants are inhabitants of waste and neglected places, but are widely found. The bees work them readily obtaining both pollen and nectar. When prolific a crop of honey can be obtained; it is bright yellow and has so offensive an odour that when first extracted it is completely unpalatable. Once granulated however, the smell is lost and the honey quite good. Pollen loads are yellow. Wax cappings are coloured bright yellow, as with a number of composites.

Rhododendron (Ericaceae) There are many species in this genus grown as ornamental shrubs in Europe, but they come originally from East Asia. *R. ponticum* is native to southern Europe and Asia Minor and has become naturalized in many woods on acid soils. Bees work them readily both for honey and pollen. Cases have occurred of bee poisoning in Scotland (see p. 192). Honey from rhododendron has made some humans quite ill, though never in my experience in the British Isles.

Robinia pseudoacacia Acacia or false acacia (Papilionaceae). It is one of the main honey sources in eastern Europe and Italy. The honey is light and of delicate flavour. Where it grows and flowers well, the bees will work it.

Runner Bean *Phaseolus multiflorus* (Papilionaceae). This is the climbing or dwarf bean with the bright scarlet flowers. It requires the service of bees to pollinate it and is worked eagerly by them both for nectar and pollen. The honey is poor quality.

Sarothamnus *see* Broom

Sainfoin *Onobrychis viciifolia* (Papilionaceae). This red clover-like flower which used to be grown for sheep and horse fodder is now very difficult to find, although there are still a few acres grown. There is little doubt it produces the finest honey. The plant produces large quantities of yellow oil, brought in by the bees with the pollen, which stains the wax, the frames, the hive's internal walls and the crown boards a bright yellow. The honey itself is bright golden and of glorious taste, finely granulated. Pollen loads brown in colour. Colony density $1\frac{1}{2}$ to 2 per acre. Flowers June–July.

Sea Lavender *Limonium vulgare* (Plumbaginaceae). Where there are extensive stands of this plant on salt marshes a crop of honey can be harvested, given reasonable weather. Flowering in August and early September it is one of the last flows. The honey is a medium amber and of very good flavour.

Senecio *see* Ragwort

Sweet chestnut *Castanea sativa* (Fagaceae). This tree produces vast quantities of pollen and I suspect by the eagerness the bees work it they obtain nectar as well, certainly in some parts of Europe. The honey is reputed to be yellow and very poor quality. Pollen load colour is pale greenish-yellow.

Symphytum *see* Comfrey

Thistle *Cirsium* species (Compositae) There are a number of species of thistle, and even more daisy-type composites, most of which produce quite large quantities of nectar and pollen. Many can be grown in the garden to provide very valuable pollen late in the season.

Thyme *Thymus serpyllum* (Labiatae). Another plant of the chalk lands, often growing with marjoram (see above). The honey is again of a minty flavour, and very delicious. It is best known as a honey plant in southern Europe where it is the basis of many well known honeys, such as that from Mt Hymettus in Greece.

Traveller's Joy *see* Old man's beard

Ulex *see* Gorse

Viper's Bugloss *Echium vulgare* (Boraginaceae). A very local plant in

Britain but of considerable interest in Australia, New Zealand and Africa. Extremely highly productive of nectar and pollen. The honey is light and very good in flavour. The pollen loads are sax-blue in colour. Flowers June to September.

White Bryony *Bryonia dioca* (Cucurbitaceae). Bryony is another example of a plant which is gaining ground in hedgerows with the advent of mechanical hedge trimming. Bees work the flowers of both sexes very assiduously, obtaining both nectar and pollen. The honey is reputed to be yellow and extremely viscous, so much so that it is very difficult to extract from the comb. Pollen load colour is orange-yellow. Flowering period May to September.

Nearly all coloured flowers provide their quota of nectar and pollen for the bee to collect, exceptions being the double flowers, which have usually lost their stamens and nectaries, and some which have been hybridized, such as Russell lupins. Beekeepers, and some non-beekeepers, often ask, 'What can we grow in the garden for the bees?' There is very little you can grow in a small garden which will make a significant contribution to the colonies except early spring and late autumn pollen-producing plants. From this point of view I would suggest all beekeepers should grow, if they can, a golden pussy willow, or palm, for spring pollen and Gaillardia and Michaelmas daisies for autumn pollen. I hope that watching your own bees working your own flowers will give you the hours of satisfaction that it gives most beekeepers.

A massive crop being removed from the colonies in an apiary in Victoria, Australia.

11 The honey harvest

Harvest time is one of joy and great satisfaction when the tins or bottles are safely in the store, and the 'honey for tea' is your own. In this final chapter I am going to deal with the composition of honey, its physical and general properties, and the methods of harvesting and preparation for use, or sale, of honey.

An average analysis of honey is shown in the Table below. As will be seen, honey is basically a solution of sugars which make up some 79 per cent of its weight.

Composition of honey

18% Water
35% Glucose (Dextrose)
40% Fructose (Levulose)
4% Other sugars
3% Other substances

The part that makes honey unique is the vast mixture of substances found in the 3 per cent of 'other substances'. A breakdown of this 3 per cent is given in the next Table which shows that it includes vitamins, pigments, enzymes and various biologically active substances such as plant growth hormones, rooting compounds, choline and acetyl-choline.

Constituent parts of the 3 per cent 'other substances' in honey

About 15 organic acids including acetic, butyric, gluconic, malic, succinic

About 12 mineral elements including potassium, calcium, sulphur, chlorine, iron, etc.

About 17 free amino acids including proline, glutamic acid, lysine, etc.

About 4–7 proteins

As will be appreciated from the above honey is a conglomeration of materials, variation in the relative proportions of which can provide the permutations which makes every super of honey slightly different in colour, flavour, aroma and texture from the next.

Honey has a built-in antibacterial substance based upon the production of peroxide by an enzyme which is added by the bee. This active sterility of honey has caused it to be used for wound dressing, together with its other advantages of a complete lack of any side effects upon healthy tissue and the fact that it does not dry out.

Honey is hygroscopic, that is, it will take up water from the air thus increasing its water content, or decreasing its specific gravity, unless kept in airtight conditions. This is because it is a considerably supersaturated sugar solution. If sugar is added to water in a vessel it will dissolve but at any particular temperature there will come a time when no more will dissolve and solid sugar will be left on the bottom of the vessel. The solution will then be saturated. If the temperature of the solution is raised more sugar will dissolve, and if the temperature is lowered some of the sugar dissolved in the solution will crystallize out to a solid until the solution reaches the saturation point for the new temperature. In the period after the temperature has been lowered but before the sugar has crystallized out the solution contains more sugar than it would do at the saturation point and it is said to be 'supersaturated'. This is an important concept in the understanding of the properties of honey, bearing as it does on crystallization and viscosity.

Viscosity is the name given to the property of a fluid which causes it to flow slowly, or which resists an object falling through it. The greater the viscosity the slower the flow, and the slower a ball will fall through it. The viscosity of honey is mainly controlled by its gravity, and the lower the water content (i.e. the higher the proportions of solids dissolved in it) the greater will be the viscosity. It will rise very rapidly if the water content falls to 20 per cent and below, doubling between 20 and 18 per cent. Viscosity is also increased by the amount of colloid material in the honey. The colloids, which are probably small pieces of solid substances and large molecules and include proteins, have a similar electric charge and so repel each other. This repulsion again offers a resistance to movement and increases the viscosity, higher in dark than light honey. The extreme example of this is heather honey which has moved beyond a viscous fluid to become a gel. It is quite unlike any other honey in Europe, being not only a very stiff gel but a 'thixotropic' gel, that is if it is stirred it turns into a viscous fluid and flows moderately readily but on standing it reverts to a gel again.

Because they are highly supersaturated liquids most honeys crystallize fairly readily. Glucose is very much less soluble in water than the other major ingredient, fructose, and therefore it is the glucose which crystallizes out in most honeys and brings the solution back to the saturation point. Using the above analysis as a sample I would estimate that in 100 g. of honey about 22 g. of glucose would turn back into a solid, taking just over 2 g. of water with it in the crystal, leaving the rest of the water with the remaining glucose and all the fructose as a higher proportion of the whole. The solution between the crystals would now be about 23 per cent water, this increase being a factor which we shall look at again when discussing fermentation and which has already been mentioned when dealing with the effects of crystallized winter stores and bee dysentry.

Granulation will therefore be partially controlled by the amount of glucose supersaturation. Where this is high, for instance in oilseed rape honey, crystallization will be very rapid. On the other hand honey from *Robinia* is high in fructose and rarely crystallizes at all. However, the viscosity of the honey is another factor which will slow down the rate of crystallization by reducing the rate at which molecules of sugar migrate through the fluid to be deposited upon the growing crystals. Slow growth of crystals will produce large crystals, rapid growth fine crystals. So viscous honeys are likely to end up with a coarse granulation. Temperature will also make a considerable difference: raising the temperature will make the solution less supersaturated and less inclined to crystallize. Crystals will grow ever more slowly until at about 34–36°C (93–95°F) they cease to grow and begin to dissolve back into solution. If on the other hand the temperature is lowered the amount of supersaturation becomes greater but so does the viscosity which impedes the passage of the molecules and again crystal growth slows down. There is therefore an optimum temperature for the rapid crystallization of honey at 13–15°C (56–58°F) which will produce the finest crystals and therefore the most acceptable texture for most consumers.

The honey producer has control over crystal size if desired. Honey which is left to crystallize without any control from the beekeeper, if absolutely clean of particles of any sort and of air bubbles, will not crystallize readily, often for many years. Dust particles, pollen grains, air bubbles, the surface of the liquid or the wall of the container can all provide places where crystallization can begin and accelerate. Honey therefore needs to have something to hang on to in order to start its crystallization, and is also more likely to encounter particles and to crystallize most rapidly in a large bulk. If packed in small containers there may be a slow start to crystallization and often variation in final texture, few if any particles being present.

The whole problem of producing the required texture in honey can be solved by increasing the number of existing crystals in the solution, and the best way is to add about 5 per cent of a crystallized honey of the texture required. This process is termed 'seeding' the honey, which

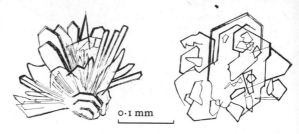

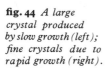

fig. 44 *A large crystal produced by slow growth (left); fine crystals due to rapid growth (right).*

0·1 mm

will rapidly crystallize to a texture similar to that of the seed, no matter what its natural crystallization would have been like had it not been seeded.

Honeys vary in colour from water white to almost black, depending upon their origins. The lighter the colour the less flavour the honey will have, and whatever subtle flavour it has will be mostly lost when crystallization has occurred. As the colour darkens so the amount of minerals and probably of proteins tends to increase the flavours present and more flavour is retained after crystallization. There is no doubt that the finest-flavoured honey is that taken straight off the hive in the comb and eaten still warm from the bees. From this time onwards the flavour is progressively lost—this happens naturally and is no fault of the beekeeper. Many of the fine flavours and the bouquet of the honey are composed of aromatic oils and other substances of plant origin which are extremely volatile and mostly lost during crystallization. This also applies to bad flavours and bouquets. Honey from ragwort (*Senecio squalidus*) is extremely offensive in smell, but once crystallized this is lost and it is as acceptable as any other honey. The fine flavours of thyme and marjoram, alas, go the same way. Heating will of course increase the rate of loss of these substances and therefore should be used as sparingly as possible during harvesting and packing.

Heating accelerates a number of the natural processes which occur all the time in honey. Two of these are used at times to monitor the amount of heating to which honey has been subjected or the length of time it has been kept before sale. They are the amount of diastase activity and the quantity of hydroxymethylfurfuraldehyde, or HMF for short. Diastase is the enzyme which breaks down starch. It is a protein and is therefore degraded by heat and by natural breakdown processes, and its quantity in honey will reduce with time and heating. Its activity is measurable and is expressed as a Diastase Number. HMF is a substance produced by the degradation of sugars in the presence of acids, and this occurs with ageing of honey and is accelerated by heating. Its presence probably causes the darkening of honey with age and heating but it is not injurious to consumers. The analysis for both diastase and HMF is complex and beyond the capability of most beekeepers, who therefore will never know for sure whether they are selling honey within the legal requirements or not. However, providing normal methods of handling honey are used and heating is kept to a minimum there should be no problem.

Removing the honey from the hive

The first thing to be done is to remove the honey from the bees, and this is called 'clearing' the supers. Clearing is accomplished using one of the following methods: shaking and brushing, using escape boards or clearer boards, using chemical repellents, blowing the bees from the supers.

Shaking and brushing is carried out as follows. The colony is smoked in the usual way and the crown board removed. The beekeeper has with him an empty super which he places on the upturned roof on the ground. The super frames are removed one at a time and shaken to get most of the bees off, the final ones being brushed off, preferably with a feather. The comb, free of bees, is then placed in the empty super. When all the frames have been 'de-beed' and are in the new super this is taken away to a place of safety or covered so that bees cannot get back into it. The now empty top super on the hive is removed to take the combs from the next hive to be cleared. This is a very quick and efficient way to remove a few supers, particularly during the season when the nectar flow is still in force. By the end of the year when the flow has ceased and bees are having to defend against robbers and wasps it can be quite exciting or even frightening for the unskilled beekeeper, and I would not recommend it to the beginner.

Clearer Boards or escape boards are the most usual method of removing bees from supers. They rely on the use of a board which allows bees to go down from the supers to the brood chamber but not to return. Two types of board are in common use, one using the Porter Bee Escape (see fig. 45) and the other using a modification of the Canadian Escape board (fig. 46). The Porter escape is a metal device in which the bees go down through the round hole, run along the metal tunnel and through the springs which prevent them from returning.

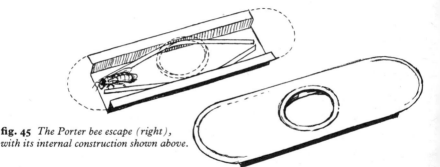

fig. 45 *The Porter bee escape (right), with its internal construction shown above.*

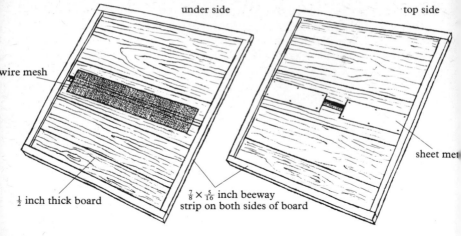

under side · top side · wire mesh · sheet metal · ½ inch thick board · $\frac{7}{8} \times \frac{5}{16}$ inch beeway strip on both sides of board

fig. 46 *A clearer board of the Canadian type.*

The springs should be kept clean and the points adjusted to about $\frac{1}{16}$ inches apart. It is possible to slide the bottom part of the device away from the top part to get at the springs. The escape suffers from the fact that drones often get stuck in them, blocking the passage of other bees through the escape, and thus most clearer boards are made with two holes to take two 'escapes', and some even three. The Canadian type escape has no moving parts and relies upon the behaviour of the bees to be effective. The bees go down through the centre hole, run along the wire gauze and down to the brood chamber through the holes at the sides. They do not return, possibly because they try to go through the gauze to the hole in the centre. Beekeepers who have put them on upside down have found that they work just as well. Drones can go down easily, there is nothing to cause jamming, nor to be propolised up, and they are much more robust for hard usage than the Porter escape, though no more efficient. It is a common practice to have crown boards with holes in the middle to take Porter escapes, the suggestion being that they then serve a double purpose. This is a fallacy, because if you take the crown board off to act as a clearer board something else has to be found to seal the top of the super and it always appears to me more sensible to have separate clearer boards kept specially for the purpose.

To clear the supers first make sure the clearer board and its escape is in the correct condition. Go to the hive, smoke the colony and remove the supers. The queen excluder can be removed, or left if more convenient, and the clearer board is put on the top of the brood chamber with the escape working in the right direction, the central hole being on top. As the clearer board is put on, the beekeeper should make sure that there is no brace comb on the top bars of the brood chamber which will block the exits of the escape. The supers are now lifted back on to the clearer board (a maximum of three to a board is the usual limit) and the beekeeper should also make sure that there is no brace comb on the underside of the bottom bars of the super combs, blocking the entrance to the escapes. The pile of supers is now examined very carefully to ensure there is no hole or crack that will allow bees or wasps to get in from the outside. It is a good idea to carry a lump of Plasticine to fill any such holes if they are found. Remember it has not mattered up to this stage if holes were present, as they would be defended by the colony, but now the bees know their honey is lost in the boxes above and the members of the colony will try to get in. If they do find a hole then their movement to and from it will attract outsiders and wasps. I have seen the best part of 40 lb. of honey lost in a couple of days in this way. The roof is then put on and the colony left for 24–48 hours.

Clearing is easier in good weather than in non-flying weather. The bees clear very readily from sealed honey but very much more reluctantly from unsealed honey and hardly at all from freshly stored pollen and brood. If after the two days are up, a lot of bees are left in the supers, often in a solid mass in one super while the others are empty, then a frame should be taken out to find what the trouble is. If it is pollen, the bees can be shaken off and the honey taken home, but should it be brood then it should be put back on the hive above a queen excluder to hatch out. It is important to find out whether there is a second queen in the colony by careful examination of the super.

Clearing with a clearer board cannot be used to remove honey from mustard, oilseed rape, kale or any of the common crucifers, as it will be crystallizing in the comb before it can be extracted and will be hard to recover. These honeys are usually taken off either by the shaking method above or the chemical repellent method below.

The use of *chemical repellents*. The desire to remove honey quickly and with only one journey has led to the use of many repellents to drive the bees from the supers. The most successful one to date is benzaldehyde, an almond-scented fluid. This is used on a board the same size as the crown board of the hive but made of soft insulation board with a half-inch beeway strip all round, or an ordinary crown board on to which is stapled a cloth. About a teaspoonful of benzaldehyde liquid is sprinkled on this as evenly as possible. The colony is then smoked, the crown board removed and the bees driven down from the top bars with smoke. It is essential to get the bees moving down in this way with smoke. The chemical-coated board is then placed on the top of the super and left for a minute or two. A brief glance will show whether the bees have gone down, and if so the super can be removed and the board placed down on the next one after smoking. In this way the supers are removed one at a time, and taken away or covered. One man can use about five boards at one time and can clear bees from supers on several hives in a short while. Should the bees only go down as far as the bottom bars of the frames the whole super can be bumped on a up-turned hive roof to knock them off. The speed of downward movement by the bees will depend upon the temperature and the strain of bee, some being repelled by the chemical much more quickly and effectively than others. The amount of benzaldehyde—a small teaspoonful to a board—will last for a whole morning's work and should not be exceeded, as too much of the repellent seems to inhibit movement entirely. The chemical is inflammable and should be kept away from flames. It oxidizes very rapidly to benzoid acid with the creation of a heat, so that if used on a cloth this should not be screwed into a ball and left lying around as there is a danger of spontaneous combustion. The great advantage of this method is that it allows one to go and clear the bees and bring home the supers in one journey, without the hard work and fuss of shake and brush; a considerable saving where one is working a number of out-apiaries. The bees are not upset by the repellent but remain quiet, tending to run and cluster, and do not get cross afterwards. The main disadvantage is that the process is much slower and more tedious in cold weather.

The use of *mechanical blowers* to remove bees from supers has been adopted widely in America. In some ways, this is the ideal method as the result is the same whatever the weather or the strain of bee. It has the advantage of the repellent and 'shake and brush' methods of being accomplished in one journey, and the main disadvantage is the cost of the equipment to do the job: possibly a home-mechanic can make his own at reasonable cost.

The basic machinery is a small petrol engine—electricity is no good for working out-apiaries—which will turn a large fan, the output of which is piped to a flexible outlet like a vacuum cleaner tube of approximately 1–3 inches diameter. The air stream should be of large volume, moving rapidly but not under high pressure. The super to be cleared is removed from the hive and the bees blown out of the super on to the ground from which they will make their own way home. Beekeepers I know who use this method appear to be quite satisfied with the result and find the bees are quiet. The experience is of such catastropic proportions to the bees—rather like driving or cutting a colony out of a tree—that they become completely disorganized and cluster in bunches for a while. It is a drastic method I would not advise for the beginner or the suburban beekeeper.

The best method for the beginner is the use of clearer boards, but if he is dealing with crucifer honey—particularly oilseed rape—then the use of benzaldehyde will be more suitable.

Honey, with typically wrinkled capping, in a brood frame, arches over a space left by the bees for the queen to lay in, even though she cannot get through the queen excluder below.

Where cells in supers are completely sealed, the honey can be taken off at any time as it should be down to the water content (20 per cent and below) that the beekeeper requires. Not so unsealed honey, however. Here it is necessary to check on the water content. Unsealed honey is unsealed sometimes because it is still being worked by the bees and has not yet reached a low enough water content for them to seal it, and sometimes because the flow of nectar has ceased and, although the honey is up to gravity, the cells are not full and so are left unsealed while the bees wait for more to arrive. We can differentiate between these two by taking out an unsealed comb of honey and holding it flat over the top of the hive, giving it a good jerk downwards towards the frame tops in the super. If no spots of honey come flying out then the honey is ready to take off and extract with the rest. If spots of liquid come out when the frame is jerked then the honey is not ready, and the super should be left on a while longer to allow the bees time to finish the job. There is never any virtue in extracting the unfinished, unsealed honey and certainly none in feeding it back to the bees.

Most beekeepers leave their honey on until the end of the season and extract in August or early September—one period of extraction with its inevitable mess is quite enough. In the rape-growing areas, however, the honey has to be removed as soon as the fields return to a green colour as the flowers fade, or the honey will become too hard to extract.

Having cleared the bees from the supers and taken these home they must be stacked in a bee-tight room or shed. The stacks should also be made bee-tight by covering top and bottom with crown boards. If bees can get at them thousands will turn up to help take the treasure home and a lot of honey can be lost, a lot of disturbance caused both to oneself, one's neighbours and to colonies in the immediate neighbourhood who will start trying to rob each other.

Decapping

This has usually been called uncapping, but it is better to use the word 'decap' as then 'uncapped honey' will not have the double meanings it has at the moment—honey which has never been capped and honey which has had the capping removed.

This is the first stage in the process of getting the honey into bottles for use or sale. The wax seal or cap has to be removed from the combs of honey which can then be put into the extractor and the honey spun out. The way in which decapping will be done will depend upon the amount of honey being handled and I would suggest the following method.

Where only half a dozen colonies are being serviced a large enamel or plastic bowl should suffice. A piece of wood is placed across the bowl with a cleat at each side to hold it steady. In the centre of the wood a large nail is driven through, point upwards. The nail should project at least the length of the lug of the frame. The frame containing the comb to be decapped is then placed on the nail, and can be revolved for easy access from any angle.

The cappings are cut off with a knife, preferably one sold for the purpose, but a sharp fluted kitchen knife will do. The fluting on the blade helps to prevent the knife being held by the viscosity of the

Decapping the honey with a steam knife. Note the method of holding the frame.

honey. The frame should be held with the top overhanging the bottom on the side being cut, so that the sheet of cappings falls away from the face of the comb and does not adhere to the honey-covered cut surface. Most people prefer to cut upwards, as this gives greatest control of the knife and a cut made with a sawing motion of the knife is less likely to cause damage to the comb than one pulled straight through by brute force. It is quite surprising how much force is needed to cut the capping off, the density of the honey having considerable bearing on the matter. If Manley super frames are used the knife is rested upon the top and bottom bars of the frame and the comb cut back level with these, any low places being decapped separately with the point of the knife. Using standard frames I try to cut about $\frac{1}{8}$ inch above the top and bottom bars to keep the comb as parallel as possible. The old idea of cutting carefully through the airspace of the cappings is a waste of time, being a slow process with no advantages in the end. The cappings drop in the bowl and can be dealt with when all the supers have been decapped. The combs, decapped on both sides, should be placed in the extractor, or on a temporary storage tray.

For those dealing with up to fifty or sixty colonies I would suggest a fairly large decapping tank constructed as in fig. 47. This is merely a

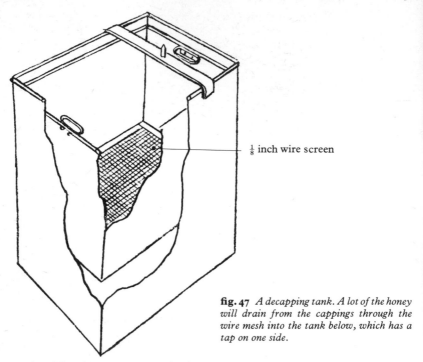

⅛ inch wire screen

fig. 47 *A decapping tank. A lot of the honey will drain from the cappings through the wire mesh into the tank below, which has a tap on one side.*

tank with a honey gate at the bottom and an internal wire-bottomed basket, the wire mesh being about eight holes to the inch. This has a metal bridge across it with a projecting pointed bolt to support the frames. Cappings fall into the basket and a large proportion of the honey drains from them into the tank. The basket should be large enough to hold all the cappings from one spell of extracting.

The use of a heated decapping tray which melts the wax is inadvisable because heating upsets the HMF level in the honey, and may cause the beekeeper to contravene the local legislation. Large honey producers may use a steam knife, but there is rarely need for speed because the process is regulated by the time of extracting rather than decapping. The use of hot water to warm the ordinary knife is a waste of time, for the knife will be cold before it has cut a couple of inches into the comb.

A stand of some sort is necessary to put the decapped combs on, and a drip tray to catch the dribbles of honey which will run from the combs. Patent devices are marketed for this purpose. During the process of decapping the combs it is absolutely necessary to prevent the cappings from getting all over the place. Some are bound to fall outside the bowl or basket. Any which fall on the floor should be mopped up immediately or they will be picked up on the shoes and carried about. Unless absolute hygiene reigns the whole house or shed will rapidly become covered with a thin layer of sticky honey. I have known beginners give up beekeeping because of the mess they got into when extracting. Chaos is not inevitable if precautions are taken and constant control maintained.

When extracting is finished the cappings have to be cleared up, and the bees can be called in to help. If you have a Miller feeder, the bees can get under the inside wall into the main body of the feeder when the syrup level has fallen to the bottom, therefore the feeder can be filled with cappings and given to the bees. Wire gauze-bottomed boxes can be made to fit inside empty supers to hold cappings, or the cappings simply fed to the bees in bowl inside a super. However they are returned to the colony, the bees will turn these cappings over and over until they are completely dry of honey and can be removed for melting and the recovery of the wax.

Alternatively, mead-makers can wash the cappings in water and adjust the density of the liquid with a hydrometer, adding water or honey until a correct mead must is obtained, before setting about making mead in their usual way. Or cappings can be dried out by centrifugal force if you have a mechanized extractor with a perforated spinning cage.

Extracting

Combs which have been decapped are put into an extractor which operates by using centrifugal force to throw honey from the comb as water is expelled from clothes in a spin drier.

Extractors, especially if made of stainless steel, are not cheap, and will usually cost at least as much as a new beehive. Beginners reluctant to spend so much before they have got the feel of the craft should contact their local beekeeping society and see if they have one which may be hired. If there is not, they could ask other beekeepers for the opportunity to borrow or hire an extractor for a day.

There are two main types of extractor: the tangential and the radial extractor as shown on page 246. The tangential type can be made with a much smaller barrel, or tank, than the radial and is therefore cheaper and more likely to be used by the beekeeper with a small number of colonies. The radial extractor is, however, the most efficient in terms of time taken and ease of use and is almost always the type used in motorized units.

The *tangential extractor*, as illustrated, is designed to hold either two super frames or two brood frames. To use this extractor the decapped combs are placed inside, resting against the cage, which supports the comb and prevents it from being torn from the frame when the cage is revolved. The handle is turned and the cage is gradually speeded up until honey can be heard, and seen, pattering

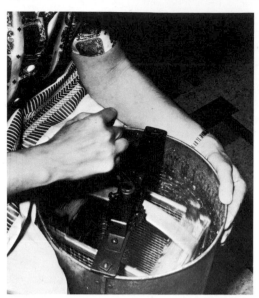

Left *A small tangential extractor suitable for a beekeeper with two or three hives.* **Below** *A motorized radial extractor being filled with frames.*

against the barrel of the extractor. Keep a steady speed until the pattering begins to diminish. The cage should then be stopped and the combs taken out and turned around so that the other face is outwards. As honeycomb has cells on each side of the central septum this new outside face will still be completely full. It is for this reason that the cage must not be speeded up too much or the weight of this honey on the inside will squash the comb against the cage, breaking the comb. Once the combs have been turned round the cage is revolved again and slowly speeded up until honey begins to patter out, this time the speed of rotation can be increased as the amount of honey being thrown out on to the barrel gets less. The speed should be increased until no more honey comes out. The cage is then stopped again, and the combs once more turned around and dried out on the other side. The now-empty combs should be taken out and replaced in their supers.

The *radial extractor* is illustrated below left. In this version the combs are placed in the slots provided like the spokes of a wheel, radiating from the centre. The drum is turned and slowly the rotor is speeded up until honey is heard pattering on the barrel. With an extractor of this type both sides of the combs are being extracted at the same time. It is possible, therefore, to increase speed gradually, keeping the honey pattering out, until the combs are dry and the job is finished. The main thing is to keep the rate of acceleration low or combs may be thrown out of their frames, particularly when they are new combs of the current season. Young white combs of this age are always fragile and should be gently used. It is important also to balance the weight of the combs in the rotor as much as possible when putting them into the extractor or it may be difficult to hold the extractor down and to stop it moving about all over the floor. Hand radial extractors usually take eight to twelve frames and motorized ones twenty or more.

Most extractors of either type have a space for retaining a fairly large amount of honey below the rotor or cage and only need emptying every three to four loadings. Nevertheless, it is often helpful to mount them on a sturdy stand which will raise the honey gate, or tap, to a level where a bucket can be stood below it to be filled with honey and then lifted to the straining tank.

Straining

The beekeeper with a very small number of colonies may let his honey settle in the bottom of the extractor in a warm room, leave it overnight, and then run it off directly into containers for use.

The beekeeper with a larger amount of honey to deal with, and particularly one who is going to sell a proportion of his honey, should pass it through a separate tank. This can be a tank of any type, made of tin plate, stainless steel or plastic. The honey can be run out of the extractor into the tank through a tap strainer which will take out most of the bits and pieces. The tin of honey is warmed quickly to about 35°C (95°F) and the honey is then poured through a cloth strainer in the honey tank. The straining cloth should be about 54 mesh to 1 inch and nylon is quicker in use than cotton. The cloth should be allowed to be low in the tank so that the honey can fill up the area around it quickly and so reduce the amount of air incorporated in the honey as it drips from the underside of the cloth. A long piece of cloth can be progressively pulled across the tank as an area becomes choked.

This sort of straining is efficient where there is no crystallization of honey in the combs. Some crystallization can escape notice, and it does not necessarily prevent the honey being spun from the combs, but it will clog the cloths very quickly and straining then becomes far too difficult and time-consuming a labour. There are two ways of getting over the problem: the honey can be heated sufficiently to get rid of the incipient crystallization or it can be left unstrained and a settling method used to remove the bits of wax and bee. I would recommend the latter method as being the best for the conservation of the aroma and flavour of the honey.

If you wish to heat the tank, it can be wound around with a flexible heating element such as is found in electric blankets or bought as pipe-lagging cable. By experimentation the amount of heat applied to the tank can be adjusted to keep the honey at about 32–33°C (90–91°F) for about a day to clear the honey. If the honey is left for a further couple of days, and the top froth is carefully skimmed off, the honey is beautifully clean and ready for packing.

Storing honey

Honey which is bottled direct from the settling tank has two faults which lower its value in most customers' eyes. Firstly a good honey will set rock hard when it first crystallizes; secondly, the honey will 'frost', shrinking away from the shoulder of the jar and showing a white, cloudy area which is often mistakenly thought to be deterioration or fermentation. These two faults in no way alter the value of the honey by reducing its flavour or its food value; they are purely faults of presentation. To provide a honey that can be removed from the jar and spread easily, and which will not 'frost' under normal circumstances, it should be removed from the settling tank into tins and stored in them until it crystallizes.

Honey tins to contain 28 lb. of honey can be obtained from the equipment factors. These are well lacquered to prevent the honey touching the iron, for if it does it will react to form a black iron tannate with an extremely bad taste, a little of which can spoil a lot of honey. These honey tins are rather expensive and many beekeepers use improvised tins of other kinds: any clean tin will do if a polythene bag is put inside to contain the honey and, when filled, closed with an elastic band. Honey keeps in this pack better than any other way and there is no chance of its reacting with the tin. Neither is there a problem of washing up the tin after the honey is gone—only the polythene bag needs replacing.

Honey should be stored at about 16–18°C (60–65°F) to get the crystallization over rapidly. After this has been accomplished the temperature of storage should be below 10°C (50°F) to prevent fermentation. Details of fermentation are given on page 253. It should preferably be used within twelve months after extraction. Longer storage increases the chance of a high HMF figure; this does not in my opinion make it any less valuable from the nutritional point of view, but it might cause legal problems.

The storage allows the first hard crystallization to occur and the initial frosting to take place. This cloudiness is partly air coming out of solution in the honey, and partly a change in the type of crystals. Crystals in a frosted area are much larger and coarser than normal crystals, and needle shaped instead of flat. The needle-shaped crystals break up the light reflected from the honey, giving the apparent whiteness. Frosted honey can be removed from the top of the tins and the rest bottled without fear of further spoilage occurring.

Warming and bottling

The normal procedure, after the honey has been allowed to crystallize in cans, is to warm it again somewhat before bottling.

A box of the sort shown in fig. 48 will warm up to 1 cwt. of honey (four standard tins) at one time. There are always hot and cold areas in such boxes, so it is advisable to move the tins around into different

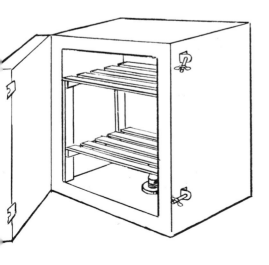

fig. 48 *An insulated warming cabinet suitable for the small-scale beekeeper who packs and sells his own honey. Made of three-ply wood with polystyrene insulation, a convenient size is 27 inches high, 22 inches wide and 20 inches deep (inside dimensions). The internal structure allows for large tins on the central shelf, and can be removed. Heating is from two 40 watt light bulbs, but other heat sources can be used provided the temperature can be carefully controlled.*

positions each day. For a larger volume it will be necessary to incorporate a fan in the design, and with an efficient fan system there should be no need to move the honey about during the warming period.

Honey may be packed for use or sale either as crystallized, or 'set' honey, or as clear honey, and these varieties will require different warming temperatures to prepare them for bottling. A fairly low temperature of 32°C (90°F) applied for 2–5 days will warm crystallized honey through with very little melting of the crystals but will bring it to a consistency which will allow it to be easily and quickly bottled using the normal tap or honey gate in a small tank. The time suggested above is for honey stored in 28 lb. lots, and will have to be increased for larger volumes and decreased for smaller ones. The variation in time is also dependent upon the hardness of the honey, which will itself depend upon its origin. A good white clover honey can seem to be almost as hard as glass, and will still be solid at the end of 4 days warming. It is, however, warm throughout and can be stirred to break up the crystals. Once this has been done it will flow readily. Other honeys such as red clover, crucifer and tree honey will only take 3 days, and will not usually need stirring. Honeydew and some dark honeys will be ready in 2 days. The beekeeper has to get to know the honeys of his area and treat them accordingly, putting the hard ones in to warm before the soft ones if he is producing a blend.

This method is dependent upon having honey which has crystallized with an acceptable texture when it first sets. If the beekeeper has honey which is coarse, and of a gravel-like texture, this can be brought right back to a fluid using the temperature suggested for clear honey, and then seeded with some honey of the right texture. If the beekeeper studies his honey and sees coarse honey turning up regularly, and can identify the source, this should be 'seeded' when it is taken from the settling tank into the cans for storage. In this way he can avoid coarse honey and the problems it may cause at bottling time.

For the production of clear honey the crystallized crop has to be rendered back to a fluid. This is usually done by heating to 52°C (125°F) for 2 days. Again adjustment will be needed for size of storage container and hardness of honey. When the honey is taken from the warming cabinet it can be strained very easily and quickly through a nylon cloth to remove from settled honey the last few bits of wax and aggregated lumps of pollen which otherwise give the final honey a cloudy appearance instead of a bright sparkle. A temperature of 52°C will still leave a considerable number of crystals small enough to get through the straining cloth, so that the honey will rapidly recrystallize, and there would hardly be time to get it to the shops and sell it before it was half set again. To avoid this, it should be heated again after bottling, this time to 62°C (145°F) for an hour in a waterbath.

This heating is done with the lids on and screwed down; there is no danger of the bottles bursting as the lids are not totally air tight. This process will give a shelf life of about 6–9 months before the honey begins to regranulate. There is no way in which clear honey can be packed on a small scale for the general market without using heat.

Regulations of sale and labelling

Regulations vary from country to country, and are updated from time to time. It is as well to check with your food authority so that you are aware of current requirements.

Labelling at present must show the name and address of the producer or packer, a declaration of the net weight and a description of the substance in the container. The description of the goods must be a correct one and must not mislead the buyer. It is important to select the design on the label with great care. For instance it is illegal to use a label showing apple blossom or an orchard if the honey in the jar did not come from apple, this being regarded as misleading information, although not conveyed in words. There may also be a problem with honeydew honey which may be required to be labelled as 'honeydew'. This is a problem here because the amount of honeydew in honey can range from hardly any to almost pure honeydew. It is difficult to draw a line between what may be labelled as honey and that which must be labelled as honeydew. Many beekeepers sell honey labelled with the name of the county of production. Details of any new regulations can be sought from the Beekeeping Associations or the Bee Press (see page 258).

Comb honey

Many beekeepers in the past used to harvest their honey in 'sections': the square wooden frames $4\frac{1}{4} \times 4\frac{1}{4}$ inches which were filled with comb and honey by the bees and sold in that form after a little cleaning up and packaging. This practice has been very much reduced of late, partly because sections take a heavy toll of whatever forage is available, since the bee consumes honey for energy in order to secrete the wax. For sections to be economically worthwhile, a heavy flow of nectar is necessary and this flow must be fairly sure each year. The honey, too, must be of high value and not of a sort which will rapidly granulate in the comb. The use of sections has been reduced in company with the great reduction in the amount of clover grown, and in inverse proportion to the increase of crucifer honey, which crystallizes so readily. Many bees are for some reason reluctant to work sections at all, and nothing will induce certain bees to work in these little square boxes. As there is always a ready sale for good comb honey, the gap has been filled in the last few years by the production of 'cut comb honey'

Three sections. The top one contains undrawn foundation (with the points of the cells at the top). A partially filled comb is below left and capped honey ready for sale is on the right.

which is ordinary well-filled super comb, cut up into about $\frac{1}{2}$ lb. pieces after removing the wire, and packed in small plastic containers. This is fairly successful in those areas where the honey does not crystallize quickly, and is of particular interest in the heather areas where the honey lends itself to this type of packing.

Heather honey

As already mentioned, heather honey from ling is quite different from any other kind and is usually obtained by taking the bees to the hills when the heather is in bloom. As heather (*Calluna*) blooms from August to the end of September according to latitude, heather honey requires a rather specialized form of management.

If the colony has already worked normal summer lowland flora it is asking a lot to expect it to carry on as late in the season as the heather flow will demand. Most queens who have worked through the year, certainly approaching two years old, will shut down their laying in August and the brood will all have emerged by the end of the heather flow. The result is that the brood chamber will be full of heather honey and tired, aged bees will show poor winter survival. One method which has been used to overcome this problem is to make up nuclei with young just-mated queens about the beginning to middle of July. These are built up and should be active on about five good combs of brood by the time colonies are taken to the heather. About ten days before going to the heather each nucleus should be united with a colony the queen of which has been removed. The brood chamber is made up solid with brood and arranged, just before moving the hive, with the sealed brood in the centre and unsealed brood on both sides. With a colony of this sort the young queen will continue laying to almost the end of September, later in some years, and the original unsealed brood will have been in heather country for about 10–21 days before it has all emerged, thus keeping the honey out of the broodnest and up in the supers.

Movement to the heather should be made when the first flowers are just coming out. The time of the main flush of nectar is uncertain; it may be right at the beginning of flowering, in the middle of the flowering period or right at the end. Examination of colonies should be made while they are on the moors so that extra super room may be given if necessary. Two supers may be filled by a really big colony.

Extracting is a problem because the honey is a jelly and will not spin out of combs in the normal way. The jelly is thixotropic, and thus if it is stirred it becomes a fluid and can be extracted normally. A form of stirring can be done in the comb using an implement which looks like a scrubbing brush set with fine steel needles for bristles. After decapping, the needles are pushed through the comb and waggled up and down quickly, and the comb is then put into the extractor and the honey spun out. For beekeepers with large amounts of comb to extract a mechanical stirrer can be obtained from Scandinavia.

The other, more traditional, method is to remove the honey and cell walls from the foundation or press the whole combs. Various heather presses are on the market and method of use is obvious: the combs or the scrapings are wrapped in straining cloths and pressed. The frames of foundation which have been scraped should be put through the ordinary centrifugal extractor as quite a lot of honey is still left on them. It is a slow process: a man working all day will have to work hard to get through more than 2 or 3 cwt. of honey.

The honey should be canned and heated for a couple of days to 40°C (115°F) before bottling. A good stir at this stage will increase the rate of flow considerably. Heather honey is ideal for the production of cut-comb honey as it does not crystallize for some while, and only then if it has some ordinary floral honey mixed with it.

Fermentation

The main process which spoils honey is fermentation. It is the consuming of the sugars of honey by yeasts which grow in size and number, using the sugars as their source of energy. When the yeasts do this they also produce many by-products which spoil the flavour and aroma of the honey. Yeasts are brought in by the bee in the nectar, their normal habitat being the nectaries of flowers. Many die when the concentration of the sugars is raised as the nectar is changed to honey but a few may survive, and these will build up a destructive population if conditions are favourable.

Honey which contains less than about 20–21 per cent water will not ferment, as the concentration of sugar is such that the yeast is unable to grow or reproduce. Once the honey has crystallized the fluid between the crystals is diluted by removal of solids, and rises by some 4–6 per cent in water content. This brings most crystallized honey into the range where fermentation can occur, but this is luckily held in check by the texture of the honey. A very hard honey will take much longer to reach a point where fermentation is noticeable than a soft honey.

Yeasts are inhibited from growing below the temperature of 10°C (50°F) and above that of 27°C (80°F). Fermentation can therefore be prevented by storing honey, either in bulk or bottles, below 10°C, at which temperature the production of HMF is also extremely slow. Storing above 27°C will produce darkening and increase the rate of HMF production. It is not usually difficult to store honey in cool climates below 10°C for most of the year to prevent fermentation.

Fermentation can be of three kinds in crystallized honey. The first is caused by a leakage into the container of water vapour which will be taken up by the surface of the honey, because it is hygroscopic, and will produce a thin layer of very dilute solution which ferments rapidly. The wet, dilute, layer on the surface of the honey with a wine-like smell is obvious and can be scraped from the honey leaving unfermented honey which can then be bottled normally. A second type of fermentation is where the surface of the honey heaves like baker's dough, although it remains fairly dry in appearance. Again the smell of fermentation and the lumpy surface gives it away, and again it is only the top $\frac{1}{2}$ inch which is affected and can be removed. The third type cannot usually be seen or smelled until the can is warmed for bottling. When it is being tipped to pour into the tank, large bubbles are seen and the smell of fermentation becomes noticeable. This fermentation usually extends through the tin from top to bottom and I would not use it for packing, but would heat it up to about 94°C (200°F) to kill the yeasts and use it to feed back to nuclei during the summer.

Honeydew honey rarely seems to ferment, but has another type of spoilage due to fungus rather than yeast. The effect is a frothy surface on the honey, gradually going deeper and deeper, at the same time producing a characteristic smell which reminds me of the smell in an apple store room when the fruit has been there some while. Spoiled honey at the top can be removed and the honey underneath will be perfectly all right.

An efficient solar wax extractor.

Beeswax

Beeswax is a valuable product of the honeybee and should be recovered from combs as they become too old for use in the colonies. It
can be made from brace comb, queen cells and any other comb
removed from the hive during manipulations, and also from the
cappings after extraction. A self-sufficient beekeeper will find that
quite a lot of this wax will be required to provide foundation to replace
combs which have been removed. The beekeeper can either make his
own foundation or trade in the beeswax in part payment for
commercially produced foundation. In good honey years he should
have a surplus of wax to sell, or to use to make candles, furniture
polish, face cream and many other home-made products.

Wax is best recovered from old combs by one or two methods. The
first—easier for the beekeeper with a small number of colonies—is the
solar wax extractor. This consists of a double box three to four feet
long and two feet wide externally with an insulated material,
preferably expanded polystyrene, sandwiched between the two
wooden skins. The box has a double-glazed lid and internally a metal
tray emptying into a metal removable container. The box is set, as
shown above, at an angle of about 40° from the horizontal and facing

due south. The sun will produce a temperature of 71–88°C (160–190°F) and wax, which melts at about 62°C (145°F) can be rendered down on an ordinary sunny day. The heat will also sterilize frames of such things as nosema spores and wax-moth eggs. If the combs are wrapped in fine cloth like cheesecloth the wax is strained at the same time, and the remains—the 'skins' or cocoons of the generations of bees who have been produced in the comb—are more easily removed. Cappings put into a muslin bag can be rendered down in the same way.

The second method of dealing with old comb is quicker and better for large quantities. It is the use of a steam-jacketed wax press. These are effective, but very expensive for the amateur. Other methods are very messy and do not recover as much wax. Mess is a major problem, wax being an intractable substance unless one can maintain it in a molten condition.

Other honeybee products

Propolis has had quite a market in recent years. If this continues it is well worth collecting, as it sells for about £1.50 an ounce. It should be kept in the small pieces as chipped from queen excluders, frames and hive bodies, and not rolled into a ball as this is not acceptable to the normal buyers.

Pollen finds a market at times and can fairly easily be trapped from the bees by making them walk through a screen on the way into the hive, the screen being of a size which removes the pollen from the bees' legs without damaging them. When pollen traps are used they should only be on the colonies for a part of each day, or on alternate days, to ensure that enough pollen gets through to the combs to provide the food needed for the colony. Production of *royal jelly*, another 'health food', and *bee-venom* require more specialized techniques and few amateur beekeepers will have the time to participate.

One final by-product of the bee which the beginner will quickly learn to appreciate is the human good-will which seems to be generated amongst beekeepers. In many years of living and working with people who to some extent share their lives with bees I have often noticed the remarkable generosity and friendship amongst them and can therefore warmly recommend the uncommitted to an involvement with bees and honey.

Bibliography

Practical beekeeping

Bailey, L., *Infectious Diseases of the Honey-bee*, London 1963
Grout, R. A., *The Hive and the Honey Bee*, Hamilton Illinois 1975
Laidlaw, H. H. and Eckert, J. E., *Queen Rearing*, Cambridge 1962
Smith, F. G., *Beekeeping in the Tropics* London 1965
Wedmore, E. B., *A Manual of Beekeeping*, London 1946
—— facsimile reprint, Warminster 1975

The life of the honeybee

Butler, C. G., *The World of the Honey-bee*, London 1974
Dade, H. A., *Anatomy and Dissection of the Honeybee*, London 1962
Lindauer, M., *Communication among Social Bees*, London 1961
Michener, C. D., *The Social Behaviour of Bees : a comparative study*, Harvard 1974
Ribbands, C., *The Behaviour and Social Life of Honeybees*, London 1953
Snodgrass, R. E., *The Anatomy of the Honeybee*, New York 1956
von Frisch, K., *Bees : their vision, chemical senses and language*, London 1956
—— *The Dance Language and Orientation of Bees*, London 1967

Honey and flora

Crane, E. (Ed.), *Honey : a comprehensive survey*, London 1975
Hodges, D., *The Pollen Loads of the Honey Bee*, London 1974
Howes, F. N., *Plants and Beekeeping*, London 1946
Mountain, M. F., *Trees and Shrubs valuable to Bees*, London 1975

Pollination

Free, G. B., *Insect Pollination of Crops*, New York and London 1970
Proctor and Yeo, *Pollination of Flowers*, London 1973

Ministry of Agriculture, Fisheries and Food *Bulletin* No. 9 'Beekeeping', No. 100 'Diseases of Bees', No. 134 'Honey from Hive to Market', No. 144 'Beehives', No. 206 'Swarming of Bees', Her Majesty's Stationery Office, London.
MAFF Advisory leaflets on Bees and Beekeeping.

Publications

British Bee Journal, 46 Queen Street, Geddington, Kettering, Northants,
NN14 1AZ (monthly)
Bee Craft, West Way, Copthorne, Sussex (monthly)
Bee World (monthly), *Apicultural Abstracts* (quarterly), *Journal of Apicultural
Research* (quarterly), Bee Research Association, Hill House, Chalfont St
Peter, Gerrards Cross, Bucks, SL9 0NR
The Scottish Beekeeper, (Editor) 23 Taybank Drive, Ayr, KA7 4RL (monthly)
The Welsh Beekeeper (Editor) The Woodlands, Llandrindod Wells, Powys

Associations

British Beekeepers' Association, General Secretary, 55 Chipstead Lane,
Sevenoaks, Kent, TN13 2AJ
Scottish Beekeepers' Association, General Secretary, 26 The Meadows,
Berwick-on-Tweed, Northumberland, TD15 1NY
Welsh Beekeepers' Association, General Secretary, Tyn-y-Berllan, Builth
Wells
Central Association of Beekeepers, Hon. Secretary, Long Reach, Stockbury
Valley, Sittingbourne, Kent
Bee Research Association, Hill House, Chalfont St Peter, Gerrards Cross,
Bucks SL9 0NR

National Beekeeping Advisors

Agricultural Development and Advisory Service (ADAS), Luddington
Experimental Horticultural Station, Stratford on Avon, Warwickshire;
Trawsgoed, Aberystwyth, SY23 4HT

Sources of supply

Taylors of Welwyn, Beehive Works, Welwyn, Herts, AL6 0AZ
E. H. Thorn (Beehives) Ltd, Beehive Works, Ragby, Lincs
Woodland Apiaries Ltd, Manor Farm, Farthingstone, Nr Towcester,
Northants
Steele and Brodie, Beehive Works, Wormit, Newport-on-Tay, Fife

Picture Credits

Alphabet and Image pages 67, 131, 207 top, 226 top; Heather Angel 22, 151,
217, 223; Australian Govt 118, 121, 232, 241; Bee Research Association 19
(Roailles), 20, 26 top (F. G. Vernon), 26 below right (G. H. Hewison), 28 right
(A. Watkins Collection), 36 (Roailles), 38 (H. Doering), 43 (W. Wittekindt),
46 (H. Doering), 58 (G. H. Hewison), 59 (A. Watkins Collection), 66, 80
(G. H. Hewison), 82 and 83 (E. Crane), 96 (A. Hyrefelt), 101 (*Hants and Berks
Gazette*), 106–7 (F. G. Smith), 125 (G. H. Hewison), 143 and 147 (F. G.
Smith), 149, 177 (P. S. Milne), 180 (H. Doering), 183 (Bee Craft), 187
(G. H. Hewison), 197, 201, 203, 207 below (I. Okada), 209, 243 (F. G. Smith),
246 top, 252 (G. H. Hewison); British Isles Bee Breeders Association 93;
C. G. Butler 26 below left and centre, 28 left, 33, 39, 56, 57; Canada
Information Service 89; S. Gates 55, 64, 114, 120, 129; E. Hooper 23, 74, 75,
77, 158, 173, 218, 221, 227; D. Husband 90; Natural History Photographic
Agency, Stephen Dalton 13, 21, 27, 30, 34, 41, 51, 52 top, 136, 205, 212;
Novosti Press Agency title page, 164, 246 below; A. E. Mc. R. Pearce 189, 226
left and below; Fred A. Richards 172, 174; A. Worth 52 below, 198. Line
drawings by Elizabeth Winson.

Index <small>Numbers in *italic* refer to illustrations</small>